T0074755

A Guide for Machine Vision in Quality Control

A Guide for Machine Vision in Quality Control

Sheila Anand

L. Priya

CRC Press
Taylor & Francis Group
Boca Raton London New York

CRC Press is an imprint of the
Taylor & Francis Group, an **informa** business

A CHAPMAN & HALL BOOK

CRC Press
Taylor & Francis Group
52 Vanderbilt Avenue,
New York, NY 10017

© 2020 by Taylor & Francis Group, LLC
CRC Press is an imprint of Taylor & Francis Group, an Informa business

No claim to original U.S. Government works

Printed on acid-free paper

International Standard Book Number-13: 978-0-8153-4927-3 (Hardback)

Library of Congress Control Number: 2019954966

**Visit the Taylor & Francis Web site at
http://www.taylorandfrancis.com**

**and the CRC Press Web site at
http://www.crcpress.com**

Contents

Preface

This book is written as an introductory guide to machine vision. The focus of this book is quality control, which is one of primary uses of machine vision in the Industry. Quality control applications have become simpler to implement and are increasingly cost-effective while providing the benefits of improved quality and business productivity.

The book follows a practitioner's approach to learning machine vision. It aims to help the readers build practical machine vision solutions. It can be used by students and academics, as well as by industry professionals. It is designed to serve as a textbook for teaching machine vision to undergraduate and postgraduate engineering students. Questions have been provided at the end of each chapter to enable readers to test their knowledge gain. Readers can build on their acquired expertise to gain more insight on machine vision technology and its emerging applications.

Chapter 1 introduces the concept of computer vision. The ultimate goal of computer vision is to mimic the human vision system; hence, this chapter draws a comparison between human and computer vision systems to understand their similarities and to identify the gaps that exist. This chapter also provides an overview of the evolution of computer vision and outlines its application in diverse areas. Computer vision and machine vision are similar in many aspects. This chapter compares computer and machine vision to highlight the subtle differences between the two systems and to distinguish machine vision systems that combine image processing with industrial automation.

Chapter 2 introduces the readers to the fundamentals of digital images. Machine vision is about using visual data for industrial automation. Digital images, its characteristics and the different types of digital images have been discussed. The mathematical representation of images as well as the formats for storage and transmission are presented. This chapter also covers some of the fundamental operations carried out on images and the broad steps involved in digital image processing.

Machine vision systems do not analyze scenes or objects directly. Instead, images of the objects or scenes are captured, and these images are processed or analyzed. Chapter 3 covers the various machine vision system components that are used for image capture and discusses how they can be integrated to obtain a working solution. Software is needed to process images and the importance of choosing the right software is discussed. The reader is introduced to automation, and some of the basic components used to build an automation solution.

Chapter 4 explores the application of machine vision for industrial quality control. The chapter starts with an overview of quality control, and traces the growth of quality control in industry. The benefits of using machine vision for quality control are elaborated, and applications of machine vision for quality control in various industries are presented. The quality control applications have been broadly classified under certain types, and examples are provided for each of these classifications.

Chapter 5 focuses on the next important step in building solutions, that is, digital image processing. Digital image processing is a vast area that covers many methods, techniques and algorithms. This chapter discusses some of the methods commonly used in machine vision systems for developing quality control applications. Digital image processing is discussed under the broad headings outlined in Chapter 2.

Building machine vision applications is a challenging task as each application is unique, with its own requirements and desired outcomes. Chapter 6 uses real-world case studies from the quality control domain to explain the process of building practical applications. The case studies begin with the requirements specifications and goes on to discuss in detail the process of selecting machine vision components. Case studies help to reinforce the concepts while providing a practical exposure to the solution building process.

No book is complete without a discussion of future trends. Chapter 7, the concluding chapter, discusses emerging trends in vision solutions. This chapter traces the history of Industrial Revolution to understand how industry has been transformed over the ages and recognize the shift from a predominately rural agrarian society to today's urban and industrial society. We are in the era of Industry 4.0, where manufacturing is going through a digital transformation and smart factories are going to be the "Factories of the Future." Automation would be taken to a new level by the mainstreaming of current and emerging technologies. Machines in factories will soon become independent entities that collect and analyze data and act upon the results in real time. Modern industries are poised to see the blurring of lines between computer and machine vision technologies, and the emergence of vision solutions that enable real-time decision making in diverse areas.

We gratefully acknowledge and thank the people who have made this book possible. We extend our gratitude to our publishers, CRC Press/Taylor & Francis Group, for supporting and guiding us in writing this book. We are indebted to the support provided by Rajalakshmi Engineering College, Chennai, India, and for permitting us to use the resources at its Centre of Excellence in Machine Vision. Our heartfelt thanks to our respective families for their unconditional support and wholehearted encouragement.

We dedicate this book to all whose passion for knowledge is matched by their dedication to learning.

Authors

Sheila Anand, a Doctorate in Computer Science is working as Professor in the Department of Information Technology at Rajalakshmi Engineering College, Chennai, India. She is an engineer by qualification, having done her BE in electronics and communication, ME in computer science, and PhD is in the area of information security. She had earlier worked in the Industry for over two decades and has extensive experience in the design and development of software applications. She is a Certified Information Systems Auditor (CISA) and has good exposure to systems audit of financial institutions, manufacturing and trading organizations. She is currently a recognized supervisor of Anna University and guides several PhD aspirants, several of whom have completed their doctoral degree. She has published many papers in national and international journals and is a reviewer for several journals of repute. She was the coordinator for a Government of India funded project (TIFAC) for machine vision at Rajalakshmi Engineering College. She is a senior member of Institute of Electrical and Electronics Engineers (IEEE) and Association of Computing Machinery (ACM) and is a member in many other professional institutions including Information Systems Audit and Control Association (ISACA) and Computer Society of India (CSI).

L. Priya is a PhD graduate working as Professor and Head, Department of Information Technology at Rajalakshmi Engineering College, Chennai, India. She has done her BE in electronics and communications, MTech in information technology and her PhD in the area of computer vision and image processing. She has nearly two decades of teaching experience and good exposure to consultancy and research. She has designed and developed many vision-based applications, and has successfully completed many projects for automobile, pharmaceutical and electronics industries. She has conducted several training programs in the area of machine vision.

1

Computer and Human Vision Systems

Computer vision (CV) has evolved into a major discipline that involves both cameras and computing technologies. However, like all major technologies, the beginning was albeit a small one. In the summer of 1966, Seymour Papert and his colleague Marvin Minsky of the Artificial Intelligence (AI) Laboratory at MIT were assigning the Summer Vision Project to undergraduates. The aim of the project was to build a system that could analyze a scene and identify objects in it. Marvin Minsky was said to have famously instructed a graduate student to "connect a camera to a computer and have it describe what it sees." However, as it turned out, it was easier said than done. Both Papert and Minsky went on to become pioneers in the fields of artificial intelligence and computer vision and have paved the way for considerable research in these respective fields.

Computer vision is the field of science/technology that deals with providing computers with vision; simply stated, how computers can, like humans, see and understand real-world objects. For instance, when humans see a flower, we can immediately perceive that it is a flower. How do we identify it? From the shape, color, smell, or all of it? So, would anything that has shape, color, and smell be a flower? That would not be true as there are other objects, sweets for example, that have shape, color, and smell. In other words, how is the human brain able to process the image of the flower that it sees and understand it as a flower? Computer vision, therefore, is not merely duplicating human vision but also the ability to process and interpret the image.

Computer vision systems use cameras to capture images in the real world. The captured image is then analyzed to understand the image so that appropriate decisions or suitable actions can be initiated. For example, the image may be captured and analyzed to determine the distance between the camera and a building for route mapping. In autonomous vehicle management, computer vision could be used to track the path of a vehicle to determine if it is being driven in the middle of its lane. Another typical example is determining the number of people present in a stadium for crowd management. In a totally different arena, a plate of food could be analyzed to find out the type and quantity of food present as well as its nutritive value. The possibilities are enormous and computer vision can virtually be applied to any field to initiate and perform various actions.

But our focus in this book is machine vision (MV). So, what then is machine vision? Computer vision covers the core technology of image analysis that can be used for any application. Machine vision, on the other hand, usually refers to a process of combining image analysis with industrial automation. For example, machine vision could be applied for automated inspection or robot guidance in industrial applications. Likewise, in process control, the parts, subassemblies, assemblies, or end products can be automatically inspected to find manufacturing defects. It is easy to see that there is considerable overlay between the fields of computer vision and machine vision.

1.1 The Human Eye

Computer vision tries to mimic the human vision system. So, we first need to understand how the human eye captures and processes images. Vision is one of most advanced of the human senses and the eyes are responsible for vision and nonverbal communication in the human vision system. The human eye uses a single lens to focus images onto a light-sensitive membrane called the retina. The obtained image is then sent to the brain for processing. The brain analyzes the images and initiates suitable action. Referring to our earlier example, once the brain identifies an image as a flower, the outcome could be simply to appreciate the beauty of the flower or action may be initiated to smell or pluck the flower.

We will now look at the anatomy of the human eye to learn about the different parts of the eye and understand how they work together to provide humans with vision. As the eyes are one of the most vital sense organs in the human body, they must be well protected from dust, dirt, and injuries that will affect human vision. The eye is enclosed in a socket that protects the eye from mechanical and other injuries. The socket is made up of parts of several bones of the skull to form a four-sided pyramid with the apex pointing to the back into the head. The eyeball and its functional muscles are surrounded by a layer of orbital fat that acts like a cushion and permits the eyeball to rotate smoothly around a virtual pivotal point. The outer structure of the human eye is shown in Figure 1.1.

The eye has several more layers of protection that safeguard the human vision system. **Eyelashes** protect the eye and prevent dust and other small particles like sand from entering the eye. The eyelashes are also highly sensitive and provide a warning when an object comes too close to the eye. Another notable feature is that the upper set of eyelashes curves upward, while the lower set of eyelashes curves downward to prevent the two sets of eyelashes from interlacing. In terms of vision, they regulate the eye from sunshine to a certain extent and provide the first layer of protection.

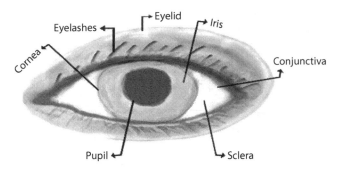

FIGURE 1.1
Outer structure of the human eye.

The next layer that protects the eye are the **eyelids**, which are movable folds of tissues. There are several muscles that help the eye to blink and help to keep our eyes open. When we blink our eye, tears are secreted, and the eyelids help to spread the tears evenly over the surface of the eye to keep it moist and lubricated. A simplified inner structure of the eye is shown in Figure 1.2.

The **sclera** is the thick white part of the eye also known as the "white of the eye." It covers the surface of the eyeball and extends all the way to the back of the eye to the optic nerve. It is opaque and fibrous and provides strength and protection to the eye. Its outer surface is covered by a thin vascular

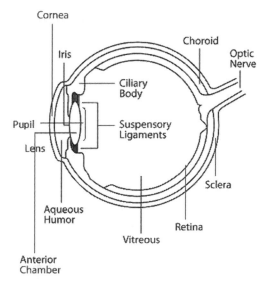

FIGURE 1.2
Inner structure of the human eye.

covering called the **episclera**. The collagen bundles in sclera are of varying sizes and are irregularly arranged; hence, the sclera is not transparent like cornea. The sclera provides a sturdy attachment for the extraocular muscles that control the movement of the eyes.

The sclera surrounds the **cornea**, which is the clear dome-shaped surface that lies in the front of the eye and covers the iris, pupil, and anterior chamber. The cornea admits and helps to focus light waves as they enter the eye. The cornea is avascular, which means that it gets no blood supply. It absorbs oxygen from the air and receives its nourishment from tears and the aqueous humor, which is a fluid in the front part of the eye that lies behind the cornea. The tissues of the cornea are arranged in three basic layers, with two thinner layers, or membranes, between them. Each of these five layers has an important function. The epithelium is the cornea's outermost layer. It acts as a barrier to prevent dust, water, and other foreign particles from entering the eye. The epithelium is filled with thousands of tiny nerve endings, which is why your eye may hurt when it is rubbed or scratched. The part of the epithelium that epithelial cells anchor and organize themselves to is called the basement membrane. The next layer behind the basement membrane of the epithelium is a transparent film of tissue called Bowman's layer, composed of protein fibers called collagen. If injured, Bowman's layer can form a scar as it heals. If these scars are large and centrally located, they may cause vision loss. Behind Bowman's layer is the stroma, which is the thickest layer of the cornea. It is composed primarily of water and collagen. Collagen gives the cornea its strength, elasticity, and form. The unique shape, arrangement, and spacing of collagen proteins are essential in producing the cornea's light-conducting transparency. Behind the stroma is Descemet's membrane, a thin but strong film of tissue that serves as a protective barrier against infection and injuries. Descemet's membrane is composed of collagen fibers that are different from those of the stroma and are made by cells in the endothelial layer of the cornea. The endothelium is the thin, innermost layer of the cornea. Endothelial cells are important in keeping the cornea clear. Normally, fluid leaks slowly from inside the eye into the stroma. The endothelium's primary task is to pump this excess fluid out of the stroma. Without this pumping action, the stroma would swell with water and become thick and opaque. In a healthy eye, a perfect balance is maintained between the fluid moving into the cornea and the fluid pumping out of the cornea. Unlike the cells in Descemet's membrane, endothelial cells that have been destroyed by disease or trauma are not repaired or replaced by the body. The cornea was one of the first organs to be successfully transplanted because it lacks blood vessels.

The **conjunctiva** is a thin, translucent membrane that lines the inside of the eyelids and covers the sclera. The conjunctiva folds over itself to allow unrestricted eyeball movement. The conjunctiva has its own nerve and blood supply. The main function of the conjunctiva is to keep the eye lubricated by producing mucus and some level of tears. It protects the eye by preventing microbes and other foreign particles from entering the eye. The **limbus** forms the border between the transparent cornea and opaque sclera.

The **pupil** is the black circle in the center of the eye, and its primary function is to monitor the amount of light that comes into the eye. When there is a lot of light, the pupil contracts to keep the light from overwhelming the eye. When there is very little light, the pupil expands so it can soak up as much light as possible. The **iris** is the colored part of the eye. The iris functions to adjust the size of the pupil. It has muscles that contract or expand depending on the amount of light the pupil needs to process images.

The **lens** lies behind the pupil and is responsible for allowing the eyes to focus on small details like words in a book. The lens is in a constant state of adjustment as it becomes thinner or thicker to accommodate the detailed input it receives. With age, the lens loses a lot of its elasticity, which often results in cataracts and other eye conditions because the lens cannot adjust as well to its surroundings as it used to.

The space between the cornea and the iris is known as the **anterior chamber**. The **posterior chamber** is the space between the iris and lens. These chambers are filled with a transparent, watery fluid known as aqueous humour, which is similar to plasma but contains low protein concentrations and is secreted from the ciliary epithelium, which is a structure that supports the lens.

The **retina** is the light-sensitive membrane that lines the inner surface of the back of the eyeball. Images are formed on the retina and it transmits those visual messages to the brain using electrical signals. **Ora serrata** is a special structure that demarcates the sensitive part of retina from its non-sensory part. This layer lies close to the choroid and consists of a single layer of cells containing the pigment. The **choroid** lies between the sclera and the retina. It supplies the blood vessels that nourish the outer two-thirds of the retina. The space between the lens and retina is covered by a transparent colorless fluid known as vitreous humor or simply as vitreous. The vitreous humor is fluid-like near the center, and gel-like near the edges. It is surrounded by a layer of collagen, called vitreous membrane, that separates it from the rest of the eye. With age, the vitreous humor begins to shrink and problems like posterior retinal detachment or retinal tears occur.

Six extraocular muscles control the movement of each eye: four rectus muscles and two oblique muscles. The medial, lateral, superior, and inferior rectus muscles move the eyeball horizontally and vertically. The superior and inferior obliques help in torsional movements like tilting the head to one side or looking up or down at an angle. The two oblique muscles of the eye are responsible for the rotation of the eye and assist the rectus muscles in their movements. The muscles of the eyes have also a role to play in the human vision system. The muscles perform a scanning function, called saccades, when looking at a large area and provide vital information to the brain.

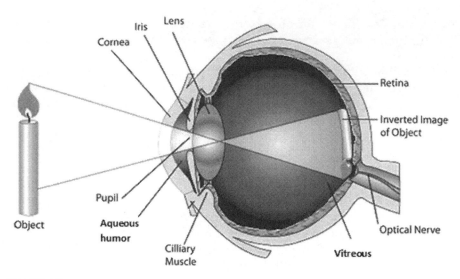

FIGURE 1.3
Image formation on the retina.

They also help in tracking moving objects in a visual field. The muscles of the eye also help in vergence, which is the simultaneous movement of both eyes to obtain or maintain a single binocular vision.

Figure 1.3 shows how the image is formed on the retina. Scattered light from the object(s) enters the eye and passes through the cornea, which bends and focuses the light on the lens. The pupil controls the amount of light that enters the eye. The lens then focuses the light onto the retina, forming an inverted image. The lens changes its shape to adjust the focus on both distant and near objects, a process known as accommodation.

Photoreceptors on the retina generate electrical signals which are sent to the brain via the optic nerve. The brain interprets the message and re-inverts the image. Hence, we see an upright image. There are two types of photoreceptors on the retina that capture the incoming light: **rods** and **cones**; these are shown in Figure 1.4.

The rods and cones are elongated retinal cells that collect the light that hits the retina. Rod photoreceptors work well in low light, provide black-and-white vision, and detect movements. Cones are responsible for color vision and work best in medium and bright light. There are three types of color-sensitive cones in the retina of the human eye, corresponding roughly to red, green, and blue detectors. The different colors are produced by various combinations of these three types of cones and their photopigments. White light is produced if the three types of cones are stimulated equally. Rods are located throughout the retina, while cones are concentrated in a small central area of the retina called the **macula**. The macula is yellow in color and absorbs excess blue and ultraviolet light that enters the eye. The macula is responsible for high resolution and therefore gives the eye the capability to discern details.

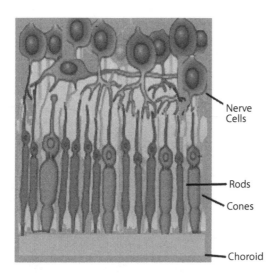

FIGURE 1.4
Photoreceptors in the human eye.

At the center of the macula is a small depression called the **fovea**, which has the largest concentration of cones. The fovea is responsible for maximum visual activity and color vision. Photoreceptor cells take light focused by the cornea and lens and convert it into chemical and nervous signals which are transported to visual centers in the brain by way of the optic nerve.

The **optic nerve** is a grouping of nerve fibers that connects the eye to the brain and transmits the visual signals to the brain. The optic nerve is mainly composed of retinal ganglion cell (RGC) axons. The optic disc is the specific region of the retina where the optic nerve and blood vessels pass through to connect to the back of the eye. This area is known as the anatomical blind spot of the eye, but it is not perceptible as the brain compensates for this. The optic nerve is a part of the eye as well as a part of the central nervous system.

It is the **brain** that constructs our visual world. The visual cortex of the brain converts the image impulses into objects that we see. However, what we see is not simply the transformation of the stimulation of the retina. The brain does a lot of work with the raw data sent by the eye to convert it to a unified, three-dimensional visual scene. Vision requires separating the foreground and background and recognizing objects presented in a wide range of orientations. The eye provides vision under conditions of varying brightness or intensity while accurately interpreting the spatial and temporal information of the objects in the scene. Human vision perception is therefore a complex process. Our intention here is to only provide a rudimentary understanding. Human vision perception continues to be the focus of much research and combines cognitive science, neuroscience, psychology, molecular biology, and many other areas.

The human vision system has two functional parts: the eye is the equivalent of a camera while the brain does the complex processing. Human eyes are known as "camera-type eyes," as they work like camera lenses focusing light onto film. The cornea and lens of the eye are analogous to the camera lens, while the retina of the eye is like the film. There are other camera-type eyes that exist in nature. Birds, for example, have camera-type eyes but have special muscles in their eyes that allow them to actively change the thickness of their lens and to alter the shape of their corneas. Whales have special hydraulics in their eyes that let them move their lenses nearer to or farther from their retinas. This unique system allows whales to see both in and out of water, and to compensate for the increased pressure they experience when they dive. Though scientists have long known how each component of a camera-type eye works, they are still a long way from being able to create a fully functional artificial eye.

The other type of eye is the compound eye that is found in insects and arthropods. Compound eyes are made up of many individual lenses. In dragonflies, for example, a single compound eye can have as many as 10,000 lenses. Some compound eyes process an image in parallel, with each lens sending its own signal to the insect's or arthropod's brain. This allows for fast motion detection and image recognition, which is one reason why flies are so hard to swat. New micro-machining technology is allowing researchers to produce tiny artificial compound eyes that mimic those found in insects. Researchers have even managed to arrange the individual lenses around a dome, which may one day be used to create devices that can see in a 360-degree angle.

Scientists are now probing nature's vision systems at the molecular level to see if they can figure out how animals get around key engineering problems. Current infrared sensors, for example, can see more than human eyes can, but they require a sophisticated cooling system to work. However, insects have developed infrared eyes without the need for such a system.

1.2 Computer versus Human Vision Systems

Humans receive a high percentage of their sensory inputs through their visual system; the other sensory inputs being hearing, touch, smell and taste. Eyeglasses, binoculars, telescopes, radar, infrared sensors, and photomultipliers are used to artificially enhance human vision to improve our view of the world. There are even telescopes in orbit, which can be termed as eyes outside the atmosphere, and many of those "see" in other spectra: infrared, ultraviolet, X-rays, etc. These give views that could not have been imagined even a few years ago, and the computer is often used to create images from these devices to enable it to be seen with our human eye.

Both human and computer vision systems convert light into useful signals from which accurate models of the physical world can be constructed. Human vision systems use the eye to capture images and the human brain to process images. Computer vision systems use cameras to capture images, while computers and image processing software are used to process images.

One difference between the two vision systems lies in the relative field of view. Human field of view is about 220 degrees. The head can rotate to include more viewing area. What we see is only at the front; we cannot see from the back of our eyes. Figure 1.5 shows the field of view of the human eye.

Computer vision systems can be built to cover the full 360 degrees field of view. Computer vision technology is mostly uniform across all parts of the field of view. In human vision, however, what we see varies across the field of view. Humans, for instance, tend to see color better at the center of the visual field. Conversely, we can detect low-light objects better at the periphery.

We saw earlier that the eye has two types of light sensor. Rods and cones are these photoreceptors that help in automatic focal adjustment of vision and an understanding of color. However, computer vision sensors presently do not have this type of specialization.

Another important key difference lies is in how human and computer vision systems transmit the image signal. In humans, nerve impulses carry the image signal from the retina to the brain. Neurons send messages electrochemically in a chain of chemical events involving sodium and potassium ions. In computer vision the signal is transmitted through electric impulses. Computer vision transmission can be much faster than that of human vision.

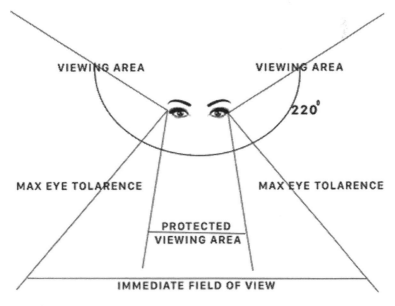

FIGURE 1.5
Field of view of human eyes.

The next difference is in where the interpretation takes place. In computer vision, the role of camera is generally to capture the image while interpretation is done using computers or embedded image processing software. Strangely in human vision systems the first stages of interpretation take place on the retina. Color and edge detection happen through ganglion cells on the retina.

Another important difference is the relation between the human eye and the brain. The human brain is deeply wired for vision, to an extent that is not currently true in computer vision technology. Some amazing experiments have shown that people who are medically blind can navigate around objects without seeing them. This phenomenon is known as blind sight. The explanation is that even when a blind person's visual processing systems does not work; the attention-directing part of the brain can still get a feed from the retina. So, while the visual processing part of the brain does not think it is processing anything, the attention-directing part of the brain knows something is there. The fact that human vision can work even for people with profoundly impaired visual systems shows how interconnected the human eye is with the wider workings of the brain. Computer vision technologies have no equivalent of blind sight, but we can try to draw a parallel. Blind humans, in addition to keeping some visual function, can also "rewire" parts of the unused brain for other tasks. This suggests there is a general learning mechanism in the brain. Deep learning techniques are capable of self-learning and are likely to be used extensively in many computer vision technologies. It should also be feasible to reuse technology for other recognition tasks such as speech recognition without major changes in the algorithm.

There is similarity, however, in the way human and computer vision systems understand image signals. Images are understood through comparison with a known set of references or stored images in a database. Human and computer vision systems evaluate whether the images are sufficiently similar to previously known examples of the same thing. However, there are difference in interpretation. Human vision, for instance, is extremely sensitive to other human faces. Research has shown that we are able to recognize faces even if we do not remember other details like a person's name, or where they live or work, etc. Optical illusion is another feature of the human eye. The brain displays an image of the scene which may differ from reality. This is because we experience the world with our senses.

However, while computer vision systems can process images at great speed, they cannot currently match the brain in terms of complexity of processing or vision perception. The memory of visual information stored by the human brain is huge and its unique association strategy permits a fast and adaptive interpretation of real-world objects. However, this can vary between individuals and, therefore, the perceived image content can differ from individual to individual. There may be subtle but significant differences in the image content. In recent years, deep learning techniques and development of huge databases have helped to improve the recognition process of computers. For example, computers can be trained to recognize faces by identifying facial characteristics like eyes, noses, mouths, etc.

To summarize, we can say that the goal of computer vision is achievement of full scene understanding. The first important aspect is to distill and represent information in huge databases in a manner that makes for easy retrieval and comparison. Second, the vast amount of computation needs to be carried out in real time to enable decisions to be taken on the fly.

1.3 Evolution of Computer Vision

We know that computer vision uses cameras to capture the image of an object or a real-world scene. Vision software is used to analyze and understand the image. The interpretation of the image is often followed by action, which may be manual or autonomous and either offline or in real-time. The areas of application of computer vision have expanded enormously due to advances in camera and computing technologies. Substantial advances in the theoretical foundations of computer vision methodologies have also contributed significantly to its growth.

Let us start at the very beginning where work carried out could be related to computer vision systems as it is known to us now. The Egyptian optical lens systems were invented as early as 700 BC. The pinhole camera dates back to early fifth century BC. However, it was not until around 1800 that Thomas Wedgwood tried to capture images using paper coated with silver nitrate. Louis Daguerre and others further established that a silver-plated copper plate when exposed to iodine vapor produced a coating of silver iodide on the surface. Silver iodide is light-sensitive, and an image of the scene could be replicated. In the mid-nineteenth century there were further developments. In England, Henry Talbot showed that paper impregnated with silver chloride could be used to capture images, which later became the foundation of the photographic industry. At around the same time, methods for capturing images electronically were being investigated. In the early twentieth century, Alan Archibald Campbell-Swinton experimented with the video camera. CRT beams were used to scan an image projected onto a selenium-coated metal plate. Philo Farnsworth further explored this idea and showed a working version of a video camera tube, which came to be known as the image dissector.

The growth of computer vision can be linked to developments in multiple areas such as optics, physics, chemistry, electronics, and computer and mechanical design. However, it would be a mammoth task to discuss or describe all the developments in these various fields. We have only looked

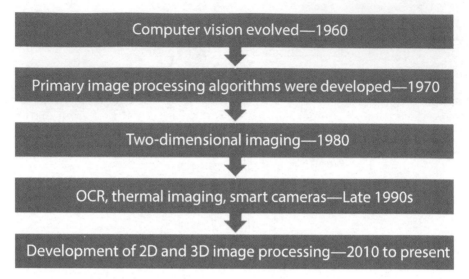

FIGURE 1.6
Evolution of computer vision.

at some significant milestones in the earlier era that can be linked to the growth and evolution of computer vision technology. The evolution of computer vision is illustrated in Figure 1.6.

The term "computer vision" evolved in the year 1960. It was originally intended to imitate the human visual system; as a stepping-stone to providing robots with intelligent behavior. Studies in the 1970s formed the basis for many of the computer vision algorithms that exist today. These were primarily image processing algorithms which included processes such as edge detection, labeling of lines, object modeling, etc.

Two-dimensional imaging for statistical pattern recognition was developed by James Gibson based on mathematical models for optical flow computation. The imaging was done on a pixel-by-pixel basis. In 1978, David Marr at the MIT AI Laboratory advocated a bottom-up approach to scene understanding. In this approach a 2D sketch is built upon by the computer to get a final 3D image. In the 1980s, the concept of optical character recognition (OCR) was developed. Smart cameras were invented in the late 1980s.

Computer vision systems continue to move forward. Computer vision growth can also be related to the development of solid-state cameras and light-emitting diode (LED) lighting which are used in machine vision system for lighting and image capture. Currently, a lot of research and commercial products are focused on 3D imaging; 3D information can be extracted from 2D data that is obtained using one or more vision cameras. Temporal information extraction can help to understand dynamic scene changes and track the motion of objects. Vision systems that scan products at high speeds have become affordable and cost-effective. Other notable areas where computer vision is being applied are in gesture-based interfaces as well as object recognition in multidimensions.

1.4 Computer/Machine Vision and Image Processing

Computer and machine vision engage techniques of image processing to understand the image or objects being viewed. Image processing techniques, in general, have an image as input and give a modified image as output. The images may be filtered, smoothed, sharpened, or even converted from color to gray scale. Computer vision and machine vision both use image processing to understand images but this is followed by interpretation to arrive at a decision or logical conclusion as output. In the case of machine vision, this decision or logical conclusion is often followed by some autonomous action in the industry.

For computer or machine vision systems, the main challenge is how to give human-like vision capabilities to a machine. Such systems should also be trained like the human brain to not only process images but also to understand and interpret images. In comparison, image processing is mostly related with the usage and application of mathematical functions and transformations over images.

In computer vision, we wish to receive both quantitative and qualitative information from visual data. Much like the process of visual reasoning of human vision, we want to distinguish between objects, classify them, sort them according to their size, and so forth. Image processing methods are harnessed for achieving tasks of computer vision. Extending beyond a single image, information in computer vision can be extracted from static images or moving images like video clips. For example, a user may want to count the number of cats passing by a certain point in the street as recorded by a video camera. Or, the user may want to measure the distance run by a player during a soccer game and extract other statistics. Therefore, temporal information plays a major role in computer vision, much like it is with our human way of understanding the world.

Machine learning and artificial intelligence methodologies are being extensively deployed to understand images. These techniques are aimed at mimicking the unique way of human reasoning. For example, a sonar machine placed to alert for intruders in oil drilling facilities at sea needs to be able to differentiate between human divers and perhaps a big fish. Sonar alone is not sufficient to detect the difference. Variables such as frequency, speed, motion pattern, and other related factors must be recorded and fed into the machine learning classifier algorithms. With training, the classifier learns to distinguish a diver from a fish. After the training set is completed, the classifier is intended to repeat the same observation as a human expert would make in a new situation. However, while we have made significant progress in this direction, we are still a long way from actually being able to replicate the human brain and its thinking or reasoning capabilities.

To gain more understanding, let us start with an overview of the image processing steps involved in computer and machine vision systems, which can be classified as image acquisition, image analysis and matching, and image understanding as shown in Figure 1.7.

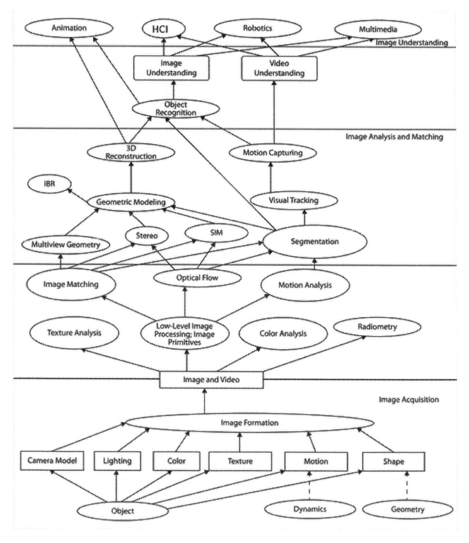

FIGURE 1.7
Computer vision—image acquisition and processing.

Image acquisition involves the use of cameras, light, lenses, and other vision accessories for appropriate acquisition of images. The cameras used for acquisition can be monochrome or color cameras as well as smart cameras. Single or multiple cameras can be used for scene capture. The focus of the vision system would be on proper acquisition of color, texture, motion, and shape of objects in the scene. The acquired images are then analyzed to extract meaningful information about the scene. Image segmentation techniques are used to partition an image into a set of nonoverlapping regions whose union is the entire image. The purpose of segmentation is to decompose the image into parts that are meaningful with respect to a specific application. Image analysis can vary depending on the type of application and the nature of the requirements.

Image understanding involves object recognition, where the objects detected in the captured image are matching with previously stored images of the objects in a database. The detected image features can also be modeled for 3D recognition and reconstruction. Once again, the methods and techniques used to solve the problems would depend on the nature of data being analyzed. For instance, different techniques can be used for different application domains, such as animation, multimedia, or robotics.

1.5 Applications of Computer Vision

Knowledge of the image processing steps in computer vision makes it easier to understand the several classes of applications that can be considered in computer vision. A well-known area of application of computer vision is in healthcare. There has been enormous progress in medical imaging techniques and modalities in the last decade. For example, ultrasound has found use in many areas that previously used X-ray or other techniques. Accompanying this progress is the greatly increased use of computer vision techniques in medical imaging. Application of computer vision in medical imaging is a challenging task because of the complexity of medical images. Medical images are processed and interpreted to extract a variety of data for diagnosis of the medical condition. These include tumor detection, malign changes, organ dimensions, blood flow in vessels, and much additional vital information. It is evident that effective use of computer vision in medical image processing can provide valuable information for diagnosis and treatment.

Another significant use of computer vision is in military applications. Automatic target recognition is one of the important applications of vision technology wherein images are automatically interpreted and analyzed to identify potential targets. This can help improve the speed and accuracy of target identification and elimination. Missile control management is another area where computer vision systems help to identify specific targets accurately to facilitate automated guidance of missiles. More advanced systems for missile guidance send the missile to an area rather than a specific target, and target selection is made when the missile reaches the area based on locally acquired image data. Modern military concepts, such as "battlefield awareness," imply that various sensors, including image sensors, provide a rich set of information about a combat scene which can be used to support strategic decisions.

Another emerging application area is in autonomous control of vehicles. Autonomous vehicles include submersibles; land-based vehicles like cars, trucks, and small robots with wheels; and aerial vehicles and unmanned aerial vehicles (UAVs). Computer vision helps the driver or pilot in navigating the vehicles and automates many tasks such as application of brakes, steering control, obstacle detection, etc. Autonomous vehicles can also be applied to specific tasks like locating and putting down forest fires, self-directed landing of aircraft, speed control, collision prevention, and many more such applications. Several car manufacturers have demonstrated systems for autonomous driving of cars and such vehicles may soon become available in the commercial market.

Computer vision is also being applied to OCR. For example, handwritten postal codes on letters can be read and automatically sorted based on address and routing destinations. Automatic number plate recognition (ANPR) is another important application that would help to identify vehicles for collecting money at toll stations as well as to aid law enforcement agencies in tracking movement of vehicles.

Computer vision technology is beginning to find a place in many consumer products, including camera phones, interfaces to games consoles, parking and driving assistance in automobiles, computer/internet image and video searches, and more recently internet shopping. It also finds application in retail for 3D model building. Visual authentication is another area of interest where computer vision can be used to automatically detect faces of humans.

Computer vision is becoming increasingly important is in the food industry, where over the last two decades, image processing has been rapidly diffused as an instrument for automatic food quality evaluation and control. Most food products have a heterogeneous matrix, therefore, their appearance properties (color, texture, shape, and size) can be strongly variable, even within the same product category. Computer vision systems can effectively replace visual inspection in different applications such as harvesting, quality control, sorting and grading, portioning, and label verification. Furthermore, they provide a more objective and standard evaluation of food quality parameters over a large number of samples.

Computer vision finds application in the entertainment and sports industry. Television and film industry use computer vision in advertising, news, drama and other applications. In sports, ball tracking and simulation are key applications of computer vision.

While there are many applications of computer vision, machine vision primarily uses computer vision in the context of industrial manufacturing processes. This could be in the inspection process itself, such as, checking a measurement or identifying whether a character string is printed correctly or through some other responsive input needed for control, such as, robot control or type verification. The machine vision system can consist of one or several cameras all capturing, interpreting, and signaling individually with a control system set to some predetermined tolerance or requirement. Manufacturers can significantly benefit by using machine vision to meet the requirements for high-quality manufactured products demanded by customers.

1.6 Summary

The study of machine vision is important not only from the research point of view but also because it is being commercially deployed in many industries, including the automobile, electronic, pharmaceutical, food processing, and medical fields.

The focus of this book is on the use of machine vision systems in the area of quality control in the Industry. This book aims to expose the readers to the practical aspects of building machine vision systems. To this view, the book does not venture to do an in-depth coverage of digital image processing but rather on the use of image processing for building machine vision systems.

The rest of the book is organized as follows. Fundamentals of digital images and digital image processing are covered in Chapters 2 and 5. Knowledge of machine vision system components is essential to build systems, and this has been covered in Chapter 3. Chapter 4 discusses quality control applications in the industry, while Chapter 6 uses real-world case studies to facilitate understanding of the process of building machine vision systems and solutions. Chapter 7 looks at the emerging scenario and the evolving path of machine vision.

Exercises

1. Compare and contrast computer vision and machine vision.
2. Draw a neat diagram of the human eye and label the parts of the eye.
3. How is the image formed in the retina in the human eye? Explain with an example.

4. What are the functions of the iris and pupil of a human eye?

5. What is the role of photoreceptors in the human eye? Compare the functionality of rods and cones.

6. Where are the anterior and posterior chambers located in the eye? What is their functionality?

7. How do the two eyes of a human work together? What are the advantages of having two eyes?

8. What are the two major functional parts of human visual system? Explain.

9. What is computer vision syndrome? How does it relate to the human eye?

10. What is a compound eye? What are the advantages of compound eye? Name any two insects with compound eye and explain how they use the compound eye?

11. What are the similarities and the differences between the human and computer vision systems?

12. Trace the evolution of computer vision expanding on the information given in this chapter.

13. What is the difference between image processing and image analysis?

14. What is Machine Learning? Explain Machine Learning with an example on image processing.

15. How is pattern recognition different from computer vision?

16. What is an UAV? Explain the role of computer vision in an UAV?

17. What is AI? Use examples to explain how AI can be used in image processing.

18. Consider an industrial application of machine vision. Prepare a detailed note on the application requirements.

19. Envisage and explain an application of computer vision for health-care of the elderly.

20. Consider any specific computer vision problem in which active research is being carried out. Discuss the challenges and the progress of research in this area.

2

Digital Image Fundamentals

The previous chapter introduced machine vision, its evolution, and its various applications. We know that machine vision applications use information from visual data for industrial automation. For example, quality control applications can be developed to determine defects in parts produced, as well as to segregate good parts from defective parts, through visual data or information obtained by capturing images of the parts. In this chapter, our focus will be the visual data, or in other words, the digital image. We will learn about basic terminologies and some fundamental operations relating to digital images.

Before the advent of digital cameras, images were captured on photographic films which were made up of light-sensitive silver halide emulsion coated on a flexible base. The film was exposed and developed using a chemical bath. Printed on photographic paper, such images are known as "analogue" images. To convert these photos to digital images, the photos can be scanned or captured using a digital camera. In general, photos, pictures, drawings, video clips, and images produced from medical equipment or automated machine output can be digitized or converted to digital images. Once converted, these digital images can be stored on hard disk or other electronic media and can also be transmitted or processed.

Machine vision systems that are used for quality control in industry typically use digital cameras to capture images. While digital image representation is common across all types of digital images, our focus would be on digital images that are captured using a machine vision camera.

2.1 Digital Image

A digital image is made up of pixels or picture elements. Each pixel represents the light intensity at a point in the digital image. Low light produces dark pixels, while bright light creates brighter pixels.

Image resolution is measured by the total number of rows (N) and columns of pixels (M). For example, an image represented with 10 rows and 10 columns would have a resolution of 100 pixels. Likewise, an image with a resolution of 640 × 480 would have 3,07,200 pixels where 640 denotes the number of columns (width) and 480 refers to the number of rows (height).

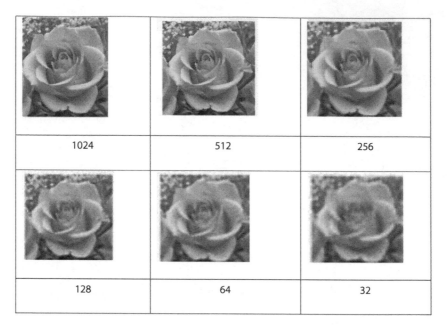

FIGURE 2.1
Images taken at different resolution.

The higher the resolution, the greater is the number of pixels used for representation of the image. In other words, high-resolution images will have more pixels in comparison to low-resolution images.

Image size is the physical dimensions of the image when it is displayed on screen or printed out. Figure 2.1 shows an image with the same size but with different resolutions.

We can see from Figure 2.1 that if image size is kept the same and the pixel count is reduced, the image detail gets reduced. Conversely, if the image size is reduced and the pixel count is kept the same, the detail of the picture remains the same as the number of pixels used to depict the image remains the same, as seen in Figure 2.2. For such displayed or printed images, the resolution detail is represented using *ppi* (pixels per inch) or *ppcm* (pixels per centimeter).

FIGURE 2.2
Images taken for different sizes with same pixel count.

2.2 Monochrome and Color Images

Digital images can be classified as binary, grayscale, or color images. Binary images are represented by using 0s and 1s where 0 represents black and 1 represents white. Monochrome refers to a single color that can have different intensity values. Black and white as well as grayscale images are commonly referred to as monochrome images.

In grayscale images, the intensity of light or gray level is represented using integers. The light intensity can vary from black to grays and finally to white. For example, 0 indicates black, while 255 represents white, giving a total of $256 = 2^8$ different levels of gray value. In other words, 8 bits are used to represent a pixel value. The number of distinct gray levels is therefore a power of 2, that is, $L = 2^B$, where B is the number of bits in the binary representation of the brightness levels L. When B>1 we speak of a gray-level image; when B = 1 we speak of a binary image. Some of the common values of B used to represent digital images are 8, 16, 32, 64, 256, etc. For example, if B = 16, the number of distinct gray levels that can be used to represent the digital image would be 65,536.

Color is a key element of visual information. The real world around us is full of colors that make it look attractive and visually appealing. Figure 2.3 shows the grayscale image of paper dolls.

When we add color to this picture, it will make the image vibrant and pleasing to the eye. We can use color to distinguish and differentiate between objects. The human visual system, we know, can recognize objects using multiple aspects such as shape, color, texture, and even the way an object moves. In Figure 2.3, for example, the doll shapes can be separated based on their color.

Traditionally, most of the image processing has been carried out using grayscale images. However, this is starting to change, and the importance of information provided by color is being recognized as well as researched.

FIGURE 2.3
Grayscale image of paper dolls.

For example, color information is being used to demark and identify the objects in various applications, such as three-dimensional scene analysis, multispectral imaging, and in security applications like video surveillance.

But first, let us understand how color images are represented. We know that sunlight is white, but water droplets can break it up into seven different colors to form the rainbow. Sir Isaac Newton in his famous experiment used a prism to break white light into a color spectrum, which he labeled as VIBGYOR—violet, indigo, blue, green, yellow, orange, red. He also proved, by placing another prism upside-down, that these colored lights can be recombined to form white light.

We saw in Chapter 1, that the human eye has three types of color photoreceptor cone cells, namely, red, green, and blue. Likewise, in digital images, three numerical components are used to describe color. The values per pixel are a measure of the respective color intensity. While these color components can be represented using different models, we shall discuss some of the commonly used ones—the RGB, CMY, YCrCb, HSV, and HSL color models.

In the **RGB color model**, three primary colors—red, green, and blue—are used to represent the color in the image. Each pixel is represented with three numerical values, the first value is the amount of red, the second is for green, and the third is for blue. The RGB color model is an additive color model based on adding and mixing light where the primary colors are combined to form other colors as shown in Figure 2.4.

It can be seen from the figure that red, green, and blue combine to form white. Absence of these primary colors would be black. Likewise, red and green combine to form yellow, red, and blue form magenta, and green and blue combine to form cyan. The intensity of these RGB colors can vary from zero to maximum value. If 8 bits are used to represent a pixel, then we have 256 shades ranging from 0 to 255. So red would be represented by (255,0,0), green by (0,255,0), and blue by (0,0,255); yellow would be (255,255,0), cyan would be (0,255,255), and magenta would be (255,0,255). A color image can be

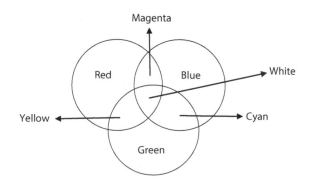

FIGURE 2.4
RGB model.

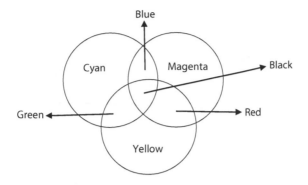

FIGURE 2.5
CMY model.

separated into its red, blue, and green components, which can then be individually converted to grayscale images.

While the RGB model is an additive model where colors are added together to form combinations, **CMY** is a subtractive model based on color pigments or inks as shown in Figure 2.5.

Here, you start with white and subtract color to get the color you want. For example, white from which red is subtracted leaves cyan. Likewise, white from which green is subtracted would be magenta. CMY colors can be obtained from the RGB model by subtracting from 255. Hence,

$$\begin{pmatrix} C \\ M \\ Y \end{pmatrix} = \begin{pmatrix} 255 \\ 255 \\ 255 \end{pmatrix} - \begin{pmatrix} R \\ G \\ B \end{pmatrix}$$

For example, if the RGB values of an image are (232,021,156), then CMY is represented as follows:

$$\begin{pmatrix} C \\ M \\ Y \end{pmatrix} = \begin{pmatrix} 255 \\ 255 \\ 255 \end{pmatrix} - \begin{pmatrix} 232 \\ 021 \\ 156 \end{pmatrix} = \begin{pmatrix} 023 \\ 234 \\ 099 \end{pmatrix}$$

CMYK model image representation is generally used for printing. In CMY, black is obtained by combining all colors. To avoid wastage of ink and also to get the jet-black coloring, black is used as a separate color to become CMYK as shown in Figure 2.6.

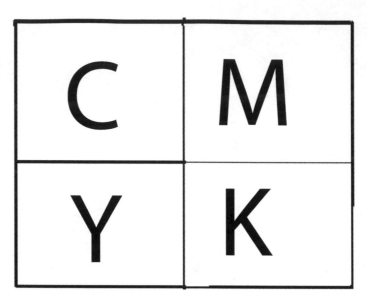

FIGURE 2.6
CMYK with black as separate color.

The **YCrCb** color model is yet another model that is used in video and digital photography systems. Y is the brightness (luminance or luma) component and Cr and Cb are the blue-difference and red-difference chroma (color) components. RGB to YCrCb conversion is obtained using the pixel intensity values of RGB, namely, E_R, E_G, and E_B. A typical conversion to luminance-chrominance is then given by

$$Ey = 0{:}299 \times Er + 0{:}587 \times Eg + 0{:}114 \times Eb$$

$$Ec_b = -0{:}169 \times Er - 0{:}331 \times Eg + 0{:}500 \times Eb$$

$$Ec_r = 0{:}500 \times Er - 0{:}419 \times Eg - 0{:}081 \times Eb$$

where Ey is between 0 and 1 and EC_b and EC_r are between -0.5 and 0.5. Conversion to 8-bit values is then done by

$$Y = 219 \times Ey + 16$$

$$C_b = 224 \times Ec_b + 128$$

$$C_r = 224 \times Ec_r + 128$$

This is illustrated further using a sample conversion. For example, let the RGB values of an image be (232,021,156), then the pixel intensities (*Er*, *Eg*, *Eb*) are obtained as (R/255, G/255, B/255). Hence,

$$Er = 232 / 255 = 0.909$$
$$Eg = 021 / 255 = 0.082$$
$$Eb = 156 / 255 = 0.611$$

Substituting the values in the equations for Ey, Ec_b, and Ec_r, the values obtained as

$$Ey = 0.299 \times 0.9 + 0.587 \times 0.08 + 0.114 \times 0.61 = 0.4039$$
$$Ec_b = -0.169 \times 0.9 - 0.331 \times 0.08 + 0{:}500 \times 0.61 = 0.1265$$
$$Ec_r = 0.500 \times 0.9 - 0.419 \times 0.08 - 0.081 \times 0.61 = 0.367$$

The values of Y, C_b, and C_r can be obtained from the values of Ey, Ec_b, and Ec_r as

$$Y = 219 \times 403 + 16 = 104$$
$$C_b = 224 \times 0.126 + 128 = 156$$
$$C_r = 224 \times 0.367 + 128 = 210$$

HSV (hue, saturation, value), or **HSB** (hue, saturation, brightness), is another color model that is widely used in computer paint and gaming applications. These models are based on how humans perceive color. Hue refers to the color, saturation refers to the amount of gray in the color, and value or brightness refers to the intensity of the color. Hue represents the color angle in degrees, which will be in the range of (0,360). Saturation and value are the intensity values in the range of (0,255). Angle values for hue are shown in Figure 2.7.

The values for hue, saturation, and value are obtained from RGB using the following equations:

$$V_{\max} = \text{Max}(R, G, B)$$

$$V_{\min} = \text{Min}(R, G, B)$$

$$S = (V_{\max} - V_{\min}) / V_{\max} \qquad (\text{if } V = 0, \text{ then } S = 0)$$

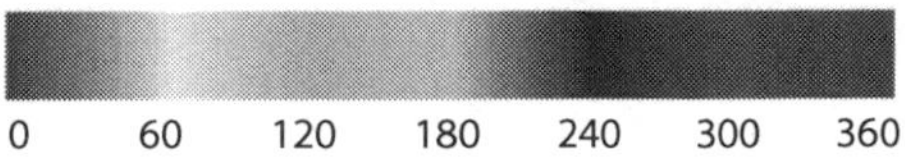

| 0 | 60 | 120 | 180 | 240 | 300 | 360 |

FIGURE 2.7
Angle values of hue.

$$H_x = 60 \times \begin{cases} 0 + (G - B) / (V_{max} - V_{min}), & \text{If } V_{max} = R \\ 2 + (B - R) / (V_{max} - V_{min}), & \text{If } V_{max} = G \\ 4 + (R - G) / (V_{max} - V_{min}), & \text{If } V_{max} = B \end{cases}$$

$$H = H_x + 360$$

For sample values of RGB (232,021,156), HSB values would be as follows:

$$V_{max} = 232$$

$$V_{min} = 021$$

$$S = (232 - 021)/232$$

$$= 0.909$$

$$H_x = -34.9 \sim -35$$

$$H = -35 + 360$$

$$= 325$$

HSL (hue, saturation, lightness) is a variation of HSV. While brightness refers to the amount of color, lightness can be explained as the amount of white or brightness relative to white. Hue is defined in the same way as in HSB/HSV. Lightness is the average of the largest and smallest RGB components and saturation depends on lightness. Higher lightness leads to more saturation.

We have looked at many models that are used to represent color. Of these, RGB is the most commonly used digital image representation. YCbCr is used in video and digital photography and in image compression. HSV separates the intensity (luma) from the color (chroma) information and is therefore useful in visualization applications, feature extractions, and illumination invariance.

2.3 Image Brightness and Contrast

We saw that intensity of light refers to the amount of light or light energy of a pixel, which is expressed as a numeric value. Brightness is a visual perception and refers to the overall lightness or darkness of an image. When brightness is high, the whitest pixels are saturated; while the blackest pixels are saturated when brightness is low. For instance, in a monochrome image, when

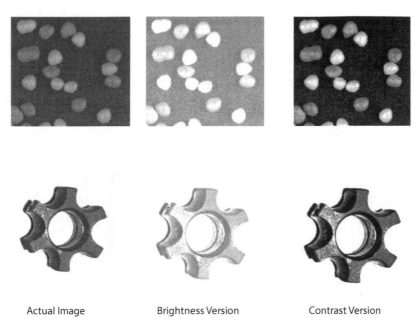

Actual Image   Brightness Version   Contrast Version

FIGURE 2.8
Actual image and its brightness and contrast version.

many of the pixels in the image have a value of 0, the image would appear dark. Conversely, when many of the pixels have a value of 1, the image would appear bright. Proper brightness is therefore important to perceive the details in an image. For example, brightness of an image can determine whether an object in the image is visible or becomes invisible as it merges with the background. That brings us to the description of contrast, which is the difference in brightness between objects or regions in the image. Figure 2.8 shows actual image and its brightness and the contrast version of the same.

Contrast separates the bright and dark areas of the image. Increasing the contrast would increase the separation between dark and bright, making shadows darker and highlights brighter. Proper contrast is important in machine vision as it enables identification of edges and boundaries of objects that appear in an image.

2.4 2D, 3D, and 4D Images

A 2D image is represented using the two dimensions of width (breadth) and height (length). For instance, a picture or drawing is a 2D image. Images that are 2D can be termed "flat" images in contrast to 3D images, which include depth. Most of the objects we see in reality, have three dimensions. For example, a car has three dimensions, whereas a picture of a car has only two dimensions. Images that are 3D can be represented using two or more 2D images along with depth map. Texture, motion, and shape are other local features that describe 3D images. A 3D image can be obtained from two or more of its 2D images by a process known as "reconstruction."

A 4D image combines 3D images with the added dimension of time; 4D and other multidimensional imaging are used in the medical field to improve the diagnostic process. For example, the basic ultrasound scan of a baby in a mother's womb is taken in black and white in 2D. Such scans are analyzed by the doctor to monitor the growth of the baby and to detect anomalies or defects in the baby. Nowadays, 3D ultrasound scans that they show pictures of the baby in three dimensions are becoming common. They provide more realistic view of the baby and parents can even identify facial features and parts of the body like legs, hand, toes, and fingers. A 4D scan shows moving 3D images of the baby like a live video effect or a movie. Hence, people can watch their baby smile or yawn!

Virtual reality and augmented reality applications are another area where multidimensional imaging is used to enhance the user viewing experience. In 3D printing, objects are fabricated using digital blueprints to build the products layer by layer. The latest in the printing technology is 4D, wherein the 3D object transforms its shape post-production. Special material is used to print the object and the transformation trigger can be water, heat, wind, and other forms of energy.

Machine vision solutions for industry are primarily built around 2D image capture and processing because 2D cameras are less expensive and images are easy to capture and process. Nowadays, 3D cameras are becoming more affordable and they provide depth, so it is possible to analyze shape and position of objects. We shall look at 2D image processing in Chapter 5 and cover 3D imaging in Chapter 7.

2.5 Digital Image Representation

An image can be considered as an array of pixels represented by an NXM matrix, where N denotes the number of rows and M the number of columns. We saw earlier that the resolution increases when the number of rows and columns, that is the total pixel count increases. An image can therefore be represented as a (NXM) matrix as shown.

$$f(x,y) = \begin{pmatrix} a_{11} & a_{12} & \cdots & a_{1m} \\ a_{21} & a_{22} & \cdots & a_{2m} \\ \vdots & & & \\ a_{n1} & a_{n2} & \cdots & a_{nm} \end{pmatrix}$$

where $\{a_{11}, a_{12} \cdots a_{nm}\}$ represents intensity values of the respective pixel points $\{(1,1), (1,2)\cdots(n, m)\}$.

The array of pixels in a 2D image can also be represented by a function $f(x, y)$, where (x, y) denotes spatial coordinates and the value of f at any point (x, y) is proportional to the brightness or gray levels of the image at that point. In digital images, the pixel vales are discretized both in spatial coordinates and brightness. Each pixel in an image is represented by a number of bits that denote the intensity of the grayscale or color depending on whether it is a monochrome or color image. A pixel could be represented using a binary or more commonly hexadecimal value.

2.6 Digital Image File Formats

Digital images represented using pixels are known as raster images or bitmap images. These images are produced by digital cameras, scanners, and other electronic devices. When viewed together as a collection of pixels, such images can provide rich details as well as the ability to edit each individual pixel. A raster image can be thought of as painting a picture using different colors which are blended using a paintbrush.

The other type of digital image format is the vector image, which is described using its contents, namely, the position and size of geometric shapes, like lines and curves, that make up the image. Graphics software can produce such images. For each line or curve, the points of the line, the equation that connects the points, and the color are all defined. Vector images can be perceived as drawing the outline of shapes, such as shape of eyes or the mouth in a human face.

Both types of images when viewed may appear similar. The difference becomes apparent when you try to scale these images. Figures 2.9 and 2.10 show the raster and vector image of the letter "C." When we try to enlarge a raster image without changing the number of pixels, the image looks unclear or blurry. When a vector image is enlarged, however, the quality remains the same because the underlying equations remain the same. The resolution is dependent on the vector shapes or objects, not the number of pixels. It is like trying to draw a circle with a larger radius with the equation of the circle remaining the same.

Raster

FIGURE 2.9
Raster image of the letter "C."

Vector

FIGURE 2.10
Vector image of the letter "C."

Raster images are, therefore, best used for non-line-art images, particularly digitized photographs, scanned artwork, or detailed graphics. The reason is that these kinds of images typically include subtle chromatic gradations, undefined lines and shapes, and complex composition. Raster images can be stored in digital media using different file formats. Some of the common formats for raster images are .BMP, .JPG, .JPEG, .GIF, .TIF, .TIFF, .PNG, .PPM, .PGM, .PBM, and .PNM. Some of the common file formats for storing a vector image are .AI, .CDR, .SVG, .EPS, and .PDF.

Stored images can be retrieved, transferred over communication lines, and of course used for image processing. The images can be stored as is or compressed. Images that are compressed occupy less storage space and can be quickly and efficiently transmitted over communication lines. The compression can be either lossy or lossless. In lossless compression, the algorithm allows the original data to be perfectly reconstructed from the compressed data. In lossy image compression, however, some of the data in the image may be lost but the file sizes are smaller. This type of compression takes advantage of the fact that some information is more important than other information. For example, while visualizing with the human eye, changes in luminance in the image are less significant than variations of hue in the image. A .JPEG image has lossy compression, while .PNG supports lossless data compression. Likewise, .SVG is lossless, while .PPM, .PBM, and .PNM are lossy.

2.7 Fundamental Image Operations

We shall now look at some of the fundamental operations that are carried out on digital images. These form the basis of digital image processing.

2.7.1 Points, Edges, and Vertices

A point is an exact position or location on an image. More specifically, in Euclidean geometry, a point is a primitive notion upon which the geometry is built that cannot be defined in terms of previously defined objects. A line or edge is bounded by two distinct end points. In a Cartesian or 2D plane, an edge can be described by linear equation in slope intercept form as given in the following equation:

$$Y = mx + C$$

where m is slope and C is the y intercept (where the graph of the line crosses the y-axis).

2D images have flat shape, such as a circle, square, rectangle, triangle, pentagon, or hexagon, as shown in Figure 2.11. Each of these 2D images have edges and where two edges meet, we have a corner. A circle has one curved side and no corner.

Common 3D shapes include cube, cylinder, sphere, cone, and pyramids, as shown in Figure 2.12. In 3D, we have the concept of faces, edges, and corners. Corners are also called vertices. A face is a large surface area that can be curved or flat. For example, a cube has 6 flat faces and 12 straight edges. A cylinder has 2 flat faces, 1 curved face, and 12 straight edges. A sphere has one curved surface and no edges or corners.

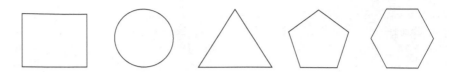

FIGURE 2.11
2D images of different shapes.

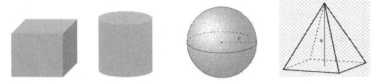

FIGURE 2.12
Common 3D shapes.

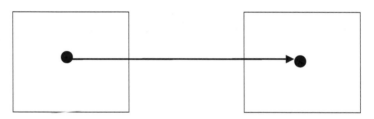

FIGURE 2.13
Point operation on pixels.

2.7.2 Point Operations

Point operations are commonly used to process digital images. The operations can be classified as point, local, and global. In point operations, the value of an output pixel depends solely on the value of the corresponding pixel in the input image, as shown in Figure 2.13.

Point operations can be used in pixel-based classification of images, often to recognize or identify images. Point operations can also be used to compare or differentiate between two or more images. The input image can be mapped onto a better output image by removing noise, which is unwanted data.

If the output value or intensity level at a pixel depends on the input intensity levels of a pixel as well as its neighboring pixels, then this operation is termed as local, as shown in Figure 2.14. Local operations can be used for a variety of processes like sharpening, blurring, color replacements, etc.

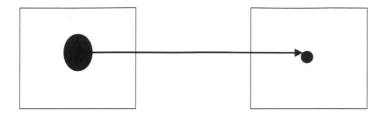

FIGURE 2.14
Local operation on pixels.

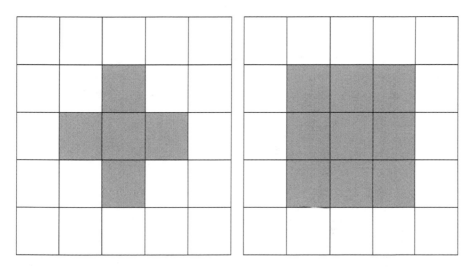

FIGURE 2.15
A 4-connected and 8-connected neighborhood.

A typical example is for edge detection, where analysis of neighboring pixels is used to detect the edge points which may lie on a straight line or curve.

When we talk about neighboring points, some of the most common neighborhoods are the 4-connected neighborhood and the 8-connected neighborhood as shown in Figure 2.15. Connectivity is used to determine whether the pixels are adjacent to one another. A pixel can have four connected pixels, which is termed 4-connectivity or 4-neighborhood. Connectivity can be extended to include diagonally adjacent pixels, which is then called 8-connectivity or 8-neighborhood. For 3D images, this can be extended to 16-connectivity.

In global point operation, the output value at a specific coordinate is dependent on all the values in the input image, as shown in Figure 2.16. Global operations include increasing/decreasing range of light and dark, brightening, darkening, color corrections, etc.

FIGURE 2.16
Global point operation.

2.7.3 Thresholding

Thresholding an image is the process of making all pixel values above a certain threshold level white, while the others are black. In a grayscale image, thresholding converts the image to a binary image. Thresholding can also be carried out on color images. Separate thresholds can be used for each of the color components, for example RGB, and the three separate images are then combined into a single image. Thresholding is used in image segmentation, which is a process that converts an image into multiple segments for easy analysis. Figure 2.17 shows a sample image and its image after the thresholding operation.

FIGURE 2.17
Sample image and its threshold version.

In simple thresholding, the same threshold value is applied to all pixels in the image. In adaptive thresholding, rather than a single value, a distribution is applied to modify the pixel values.

2.7.4 Brightness

Brightness of an image can be changed by adding or subtracting a constant value from the pixel intensity value. Figure 2.18 shows an actual image and the images obtained by adding and subtracting brightness.

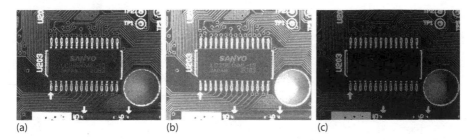

(a)   (b)   (c)

FIGURE 2.18
PCB image at different brightness levels. (a) Actual image. (b) Image after adding 20 to brightness. (c) Image after subtracting 20 from brightness.

Changing the contrast of an image can change the ranges of pixel brightness values. For instance, the subtraction and addition of 0.5 to a pixel can change the pixel intensity to gray. A value above 1.0 will increase the contrast by making bright samples brighter and dark samples darker, while a value below 1.0 will do the opposite.

2.7.5 Geometric Transformations

Geometric transformations are widely used for image registration and image correction. Image registration is used to transform different sets of data into a single coordinate system. It can be used, for instance, to align multiple scenes into a single integrated image, such as data from multiple photographs or sensors or data taken at different times or from different viewpoints. Image correction is used to adjust images for errors due to lens distortion or those that occur due to camera orientation. There are many types of transformation that can be performed on digital images. We shall restrict our discussion to a few key types of transforms that are generally used in processing digital images in machine vision applications.

2.7.6 Spatial Transformation

A spatial transformation of an image is, in general, a geometric transformation of the image coordinate system. Spatial transformation is used for image mapping. Each point (x, y) of an image A is mapped into a point (u, v) in a new coordinate system where,

$$u = f1(x, y)$$

$$v = f2(x, y)$$

The mapping of (x, y) to (u, v) is shown in Figure 2.19.

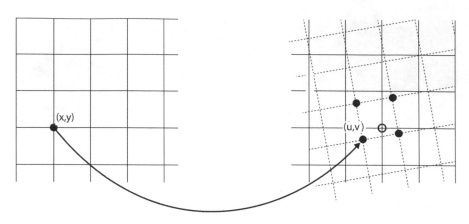

FIGURE 2.19
Mapping of (x, y) to new domain as (u, v).

During the process of mapping, an image array is mapped to discrete points in the new domain. As shown in the figure, the mapped points may not fall on grid points in the new domain.

2.7.7 Affine Transformation

An affine transformation is a geometric transformation that preserves co-linearity and ratios of distances of the points. For instance, sets of parallel lines remain parallel after an affine transformation. In other words, all points that lie on the original line will continue to lie on the same line after transformation. Likewise, the relative distance of the points in the line remain the same after transformation. For example, the midpoint of the line will remain the midpoint even after the transformation. Hence, affine transformation is a linear mapping method that preserves the points, straight lines, and planes. However, length and angle are not preserved; for example, a circle can become an ellipse.

The point (x_x, y_y) that is obtained as a result of transformation T on the point (x, y) is written as

$$(x_x, y_y) = T(x, y)$$

The inverse transform is then represented as

$$(x, y) = T^{-1}(x_x, y_y)$$

For example, consider a simple transformation that is used to stretch an image so that its width and height are double the original values.

Let $T(x,y) = (2x, 2y)$. Then,

$$T^{-1}(x_x, y_y) = \left(\frac{x_x}{2}, \frac{y_y}{2} \right)$$

Various transforms that can be performed on digital images are translation, rotation, and scaling, shearing, and projective transform. **Translation** is carried out by shifting every point in an image by a vector (d_x, d_y). Translation of (x, y) to (x', y') is represented as:

$$x' = x + d_x$$

$$y' = y + d_y$$

Conversely,

$$x = x' - d_x$$

$$y = y' - d_y$$

This could be represented in a matrix form as

$$\begin{pmatrix} x' \\ y' \end{pmatrix} = \begin{pmatrix} x \\ y \end{pmatrix} + \begin{pmatrix} d_x \\ d_y \end{pmatrix}$$

Scaling refers to the resizing of a digital image. Scaling can be upscaling or downscaling. Magnification is known as upscaling or resolution enhancement. Zoom out is known as downscaling. Upscaling of image (u, v) is represented as

$$u = x * p$$

$$v = y * q$$

where (x, y) is actual image coordinate, and p and q are the scaling factors. Conversely, downscaling of image (u, v) is represented as

$$u = x / p$$

$$v = y / q$$

where (x, y) is actual image coordinate, and p and q are the scaling factors.

Rotation is used to tilt an image. The image can be tilted by an angle α, which is represented as

$$x' = x \cos \alpha + y \sin \alpha$$

$$y' = -x \sin \alpha + y \cos \alpha$$

In matrix form,

$$\begin{pmatrix} x' \\ y' \end{pmatrix} = \begin{pmatrix} \cos \alpha & -\sin \alpha \\ \sin \alpha & -\cos \alpha \end{pmatrix} \cdot \begin{pmatrix} x \\ y \end{pmatrix}$$

Shearing is the kind of mapping that is used to displace every point horizontally, by an amount proportionally to its coordinate. In shearing, points in a specific line remain fixed while other points are shifted parallel by a distance that is proportional to their perpendicular distance from the fixed line. Shear factor can be defined as the distance a point moves due to shear divided by the perpendicular distance of a point from the invariant line.

Shearing can push an image in sideways, upward, or downward. Shearing in sideways is represented as

$$\begin{pmatrix} 1 & 1 \\ 0 & 1 \end{pmatrix}\begin{pmatrix} x \\ y \end{pmatrix} = \begin{pmatrix} x + y \\ y \end{pmatrix}$$

Similarly, upward shearing is represented as

$$\begin{pmatrix} 1 & 0 \\ 1 & 1 \end{pmatrix}\begin{pmatrix} x \\ y \end{pmatrix} = \begin{pmatrix} x \\ x + y \end{pmatrix}$$

General case for shearing in x direction is

$$\begin{pmatrix} 1 & s \\ 0 & 1 \end{pmatrix}\begin{pmatrix} x \\ y \end{pmatrix} = \begin{pmatrix} x + sy \\ y \end{pmatrix}$$

Mirroring is another operation where the vector's direction is reflected and its length remains unchanged. These vectors are mirrored through a parallel plane that passes through the origin. This is a useful property as it allows the transformation of both positional vectors and normal vectors with the same matrix.

The different types of transformations discussed above are represented diagrammatically in Figure 2.20.

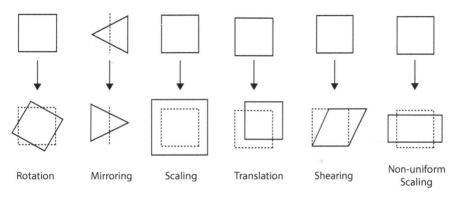

| Rotation | Mirroring | Scaling | Translation | Shearing | Non-uniform Scaling |

FIGURE 2.20
Different composites of affine transformations.

In general, an affine transformation is a composition of rotations, mirroring, scaling, translations, and shearing and can be represented as follows:

$$u = c11x + c12y + c13$$

$$v = c21x + c22y + c23$$

where (u, v) is the affine transformed image; $c13$ and $c23$ affect translations and scaling, $c11$ and $c22$ affect mirroring, and the combination affects rotations and shears.

2.7.8 Image Interpolation

The transformed images may contain some unknown point coordinates. These unknown coordinates can be filled with points based on the known values. Image interpolation is the process of finding or estimating the unknown values based on the known pixel values. For example, when an image is zoomed, the number of pixels increases, and the unknown pixel values need to be filled in. There are many types of interpolation techniques and we will discuss a few of the techniques commonly used in image processing.

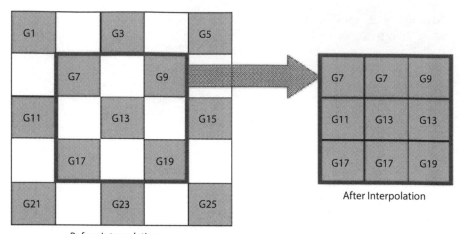

Before Interpolation

FIGURE 2.21
Nearest-neighbor algorithm.

2.7.8.1 Nearest-Neighbor Interpolation

In **nearest-neighbor interpolation**, image interpolation works in two directions and tries to achieve a best approximation of a pixel's intensity based on the values at surrounding pixels. For example, each interpolated output pixel can be assigned the value of the nearest sample point in the input image. This is represented pictorially in Figure 2.21. This interpolation can be represented using the following equation:

$$h(x) = 0 \text{ for } x > 1$$

$$= 1 \text{ for } x < 1$$

The frequency response of the nearest-neighbor kernel is

$$H(\omega) = \text{sinc } (\omega / 2)$$

This type of interpolation applies to X, Y, Z data, where the X and Y values describe arbitrary points in the Z = 0 plane. Although this method is very efficient, the quality of image is poor.

2.7.8.2 Bilinear Interpolation

When an image needs to be scaled up, each pixel of the original image needs to be moved in a certain direction based on the scale constant. However, when scaling up an image by a nonintegral scale factor, there are pixels (i.e., *holes*) that are not assigned appropriate pixel values. In this case, those *holes* should be assigned appropriate RGB or grayscale values so that the output image does not have non-valued pixels.

Bilinear interpolation can be used where perfect image transformation with pixel matching is impossible. Appropriate intensity values can be calculated and assigned to pixels. Unlike other interpolation techniques such as nearest-neighbor interpolation and bicubic interpolation, bilinear interpolation uses values of only the four nearest pixels, located in diagonal directions from a given pixel, in order to find the appropriate color intensity values of that pixel.

This algorithm in general reduces some of the visual distortion caused by resizing an image to a nonintegral zoom factor, as opposed to nearest-neighbor interpolation, which will make some pixels appear larger than others in the resized image.

2.7.8.3 Bicubic Interpolation

Bicubic interpolation is an extension of cubic interpolation for interpolating data points on a two-dimensional regular grid. The interpolated surface is smoother than corresponding surfaces obtained by bilinear interpolation or nearest-neighbor interpolation. Bicubic interpolation uses 16 nearest neighbors instead of 8 nearest neighbors of a point. The intensity value $V(x, y)$ of the unknown point from the known (x, y) is obtained using the following equation:

$$V(x,y) = \sum_{I=0}^{3} \sum_{i=0}^{3} aij \, x^i \, y^j$$

Bicubic interpolation is often chosen over bilinear or nearest-neighbor interpolation when speed is not an issue. In contrast to bilinear interpolation, which only takes 4 pixels (2 × 2) into account, bicubic interpolation considers 16 pixels (4 × 4), hence, images rendered with bicubic interpolation are smoother and have less interpolation distortion. However, analyzing 16 points takes time and this is crucial when making decisions on outputs of real-time analysis.

2.8 Fundamental Steps in Digital Image Processing

The digitized image is processed to interpret and understand the image contents. The broad steps used in processing a digital image are shown in Figure 2.22.

FIGURE 2.22
Fundamental steps in digital image processing.

Preprocessing is carried to improve the visibility of the edge points and to remove extraneous noise in an image. Image enhancement and image interpolation are some of the key techniques used in preprocessing. Image enhancement is used to bring out obscured details or highlight certain features of interest in an image. Image interpolation is a technique that is used to predict the unknown values of the pixel points. There are many types of interpolation techniques and we have earlier discussed a few of them, namely, nearest-neighbor, bilinear, and bicubic. There are many more interpolation techniques include B-spline, Lanczos, DWT, and Kriging.

Segmentation is used to subdivide an image into its constituent regions or objects for ease of processing and analysis. Image segmentation can be used to locate objects and edges. The pixels in a region would be similar with respect to certain aspects of the images such as color, intensity, or texture. Adjacent segments can be dissimilar with respect to some aspects of the image. Segmentation partitions an image in a manner that is meaningful to the application. For example, in an image of natural scenery, one of the segmented portions might contain the mountain, another a stream, valley, or the foothills of the mountain. A segmented region could also contain multiple objects that should be recognized individually. Segmentation must be done carefully to avoid fragmentation of the image. Image segmentation techniques are of many types and can be classified, based on the technique used, as threshold, region-based, or edge-based segmentation.

Segmented images are analyzed to detect and recognize the image features required by the application. Detection identifies the presence of a feature/object in the image, while object recognition identifies the detected feature or object. Object detection algorithms typically use extracted features and learning algorithms to recognize the features or instances of an object category. The object to be recognized is matched with a standard image template of the object, using image matching techniques such as template matching, model matching, and blob analysis or particle analysis. For example, if the objective of the application is to recognize bicycles that appear in an image, then the detected objects are matched with prestored images of bicycles to recognize them. Object recognition remains an ongoing research activity for reasons of occlusion and viewpoint variance, as well as improper illumination. Recognition becomes difficult when objects are occluded, that is, partially or fully hidden in the image. The objects can be found in different orientations (viewpoints), which makes it difficult to recognize them.

2.9 Summary

In this chapter we learnt about digital images and how to represent them. We looked at some of the key characteristics of images, such as resolution, size, contrast, and color. We saw that digital images can be represented as 2D, 3D, 4D, and other multiple dimensions. We learnt how to represent them using matrices and looked at different formats for storing and transmitting of the digital images.

Some of the fundamental operations on images were covered in this chapter, including point operations, thresholding, and how to add/subtract brightness in an image. We studied geometric transformations and affine transformations, which is a geometric transformation that preserves co-linearity and ratios of distances of the points. We learnt how to carry out process such as translation, scaling, rotation, shearing, and mirroring on images. We discussed how to use nearest neighbors and neighborhood for image interpolation. Finally, we looked at the broad steps used in image processing: acquisition, preprocessing, segmentation, and object detection/recognition. We shall cover image processing in greater detail in Chapter 5.

Exercises

1. Consider a color image of size 1024 × 1024. What is the transmission time if this image is transmitted across a channel of 2 Mbps?

2. What is Euclidian distance? How is it used to measure distance between two pixels in an image? Explain with the help of an example.

3. Explain with diagrams the terms (a) neighbors of a pixel, and (b) connectivity.

4. Compare and contrast bilinear and bicubic interpolation. Give an example of usage for each of the techniques.

5. Compute the pixel intensities of (Y, Cb, Cr) if RGB values of an image are (222,111,033).

6. Write the pseudo code for converting an RGB image into an HSV image.

7. Consider a machine vision system for OCR. List and explain the steps involved in image processing.

8. Compare and contrast vector and raster images. Give examples of images that you would store as vector and/or raster images.

9. What is image resolution? What is PPI?

10. What is the difference between brightness and contrast of an image?

11. What is Huffman's coding? How is it used for image compression?

12. What is image decompression? What decompression technique is used for JPEG files.

13. What are the number of bits required to store L gray levels when (a) L = 32 and (b) L = 256.

14. A digital image is represented as (N X M). L represents the discrete gray levels. Calculate the number of bits required to store the image given:

    a. M = N = 32 and L = 16

    b. N = 1600, M = 2100 and L = 256

15. What is image shrinking? Explain giving examples where images shrinking is used.

16. Write the pseudo code to convert a color image to a grayscale image. (Hint: Obtain the grayscale value by taking the average of red, green, and blue color values.)

17. Write the pseudo code to invert a black-and-white image. (Hint: Convert black pixels to white pixels and vice versa.) Give one example where inverting an image is used for processing an image.

18. Write the pseudo code to change the color balance of an image by multiplying each color by a user given input.

19. Explain an image processing technique that is used to improve the quality of images for human viewing?

20. What is object occlusion? Why is it difficult to recognize objects that are occluded?

3

Machine Vision System Components

We saw in the previous chapters that machine vision systems combine image capture with image processing to perform automated tasks in industry. Machine vision systems are essentially multidisciplinary, encompassing the fields of computer science, optics, mechanical engineering, and industrial automation. Machine vision systems have to be designed to be robust to function under demanding conditions of high speed, repeatedly performing the same task or set of tasks, and being capable of working 24/7.

Machine vision systems do not analyze a scene or object directly. Instead, images of the object or scene are captured, and these images are processed or analyzed. The captured image is analyzed to obtain desired data that is used for decision-making and/or controlling a process. The quality of the captured image, therefore, determines the accuracy of the decision process. In order words, the success of the machine vision application depends on the ability to obtain images that deliver the information which is useful or desirable for the application.

In this chapter, we shall cover the various components used in a machine vision application for image capture and related processing. It is essential that the vision components are properly chosen and integrated to obtain images of the desired quality. Deciding what components to use can appear to be a formidable task, as there is a wide range of components and image processing software to choose from. But the good news is that machine vision systems, in general, use certain basic components that are common for most applications. In this chapter, we shall discuss in detail some of the components typically used in a machine vision system.

3.1 Machine Vision System

The major components of a machine vision system include the camera, lens, light source, image processing software, and communication interfaces. Deciding the components of a machine vision system would depend on the application for which they are designed as well as the degree of automation required. For example, a quality control application can be designed with a single camera that captures static images of the objects to be inspected.

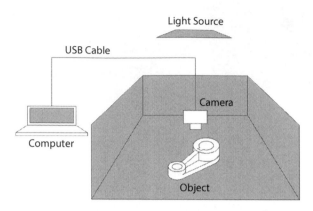

FIGURE 3.1
Machine vision system.

It would be more difficult to capture images of parts moving on a conveyer belt and the design of a robotic inspection application would be considerably more complex.

A typical machine vision system is shown in Figure 3.1. The figure shows a machine vision system that uses a single camera to capture the image of the object to be analyzed. Depending on the application, single, dual, or multiple cameras can be used for image acquisition. The sensors in the camera convert light into digital images that can be processed by a computer system. We learned in Chapter 2 that resolution of digital images is based on pixel count. The higher the resolution of the camera, the more information is captured about the image. Monochrome cameras are cheaper and provide fast and easy processing compared to color cameras. The choice of camera depends on the application for which the system is being designed.

The next component is the lens which is mounted in the camera. Lenses provide the appropriate magnification, working distance, and image resolution. Lighting is used to illuminate the image scene to improve the quality of image captured. This helps to maximize the contrasts within the image and emphasize the features that need to be analyzed.

Images are created when light strikes the object and gets reflected toward the camera. The image captured by the camera is stored in the computer through the frame grabber for further processing. Communications is concerned with interfacing the camera to the image processor and for sending the results to the automation components.

Computer systems are most commonly employed for image processing. They provide a wide variety of programmable software options which help to build code for custom processing of images. Smart cameras have a built-in processor that can be used for limited processing. The automation components are actuated or triggered by the results of image analysis.

Machine vision components are available as commercial off-the-shelf (COTS) products that can be purchased individually and assembled together to form the system. Depending on the application it is also possible to purchase an integrated system with all components. These components are discussed in greater detail in the rest of the chapter.

3.2 Machine Vision Camera

Cameras can be of two types—analog and digital. Machine vision systems predominantly use digital cameras, which can be either line scan or area scan cameras. The image sensor and the lens are key components of the camera. The resolution of the image is determined by the image sensor, while the lens determines the focal length of the camera.

3.2.1 CCD and CMOS Image Sensors

The sensor in the camera that captures the image of an object or scenery is known as the image sensor. Image sensors are divided into two types—charge-coupled device (CCD) or complementary metal-oxide semiconductor (CMOS).

Both CCD and CMOS sensors work in the same way. The sensors can be imagined as a grid of photosensitive cells (pixels). When light from the scene gets reflected toward the camera, the light falls on these photosensitive cells of the image sensor and is converted to electrical energy or charges. The difference between the two technologies lies in the way the grid is read and the digital signal is obtained.

In CCD sensors, the electrical charges are transferred line by line using control circuitry and read like a shift register as shown in Figure 3.2. The output is converted to a digital signal using an analog-to-digital converter (ADC).

In CMOS sensors, at each pixel, integrated circuits are used to convert the electrical energy to voltage, which is then transferred using wires. CMOS is illustrated in Figure 3.3.

CCD sensors produce high-quality images but consume more power and are expensive. CMOS sensors in comparison produce images of lesser quality but consume less power and are relatively inexpensive.

In CCDs, the chips are specifically manufactured or fabricated to enable the electrical charges to be transported through the chip as explained earlier. CMOS uses the regular integrated circuits manufacturing processes to create the chip and hence the cost of manufacture is less when compared to CCD sensors.

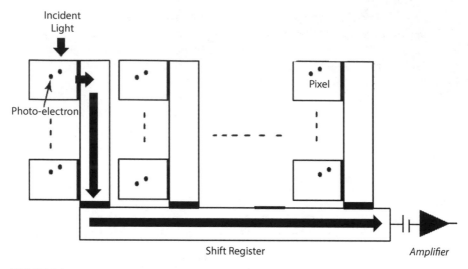

FIGURE 3.2
CCD image sensor.

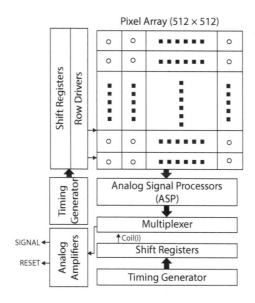

FIGURE 3.3
CMOS image sensor.

CMOS consumes less power as CCD use extra circuitry for A/D converters, etc., which require more power. The speed in CCD is limited by the transfer speed of the charge packets through the chip, while CMOS can provide higher speeds.

In CMOS sensors, voltage conversion happens at every pixel, hence the image quality is less due to higher noise. In CCDs, the conversion is done after collecting the charges from the pixels, hence the noise is less, and the image quality is better. In CMOS sensors, integrated circuits are located at each pixel, so the light sensitivity is less, as not all the light may fall on the photodiode. While this problem is not present in CCDs, the charge from one pixel can spill over to the next pixel; and this can create bright spots. The CMOS approach is, however, more flexible as each pixel can be read individually.

In summary, we can deduce that CCD sensors tend to be used in applications that require high-quality images with high pixel counts, such as in professional, medical, and scientific applications. CMOS sensors are in greater demand in consumer applications and for professional digital cameras with less exacting quality demands. However, CMOS technology is evolving and, in the future, may match CCD in producing high-quality images while retaining the advantages of less cost and smaller power requirement.

3.2.2 TDI Sensor

The time delay integration (TDI) sensor is a special kind of sensor that is used in line scan cameras. Line scan cameras scan one row of an image at a time. The image needs to be quickly recorded before the camera moves away. Hence, high light levels are required. However, this may not always possible or feasible when used in industrial settings. TDI sensors have multiple rows of pixels arranged as an array. The camera movement is synchronized with the movement of the object or scene whose image is being captured. Hence, more than one row in the sensor array captures the same image data. Therefore, it is possible to obtain a good-quality image even under low illumination conditions.

The TDI technique can be easily used with CCD technology because there is less image noise. However, it is more expensive and requires more power than CMOS technology. A newer approach is the CCD in CMOS technology. CCD-based TDI imaging is being combined with advanced CMOS drivers and readout in one single chip.

3.2.3 Camera Type

Machine vision systems essentially use two different camera types: the line scan and area scan cameras. Line scan cameras capture the image as a single row of pixels, while area scan cameras contain a matrix of pixels that capture an image of a given scene or object. The third type of camera used in machine vision applications is the smart camera, which has built-in image processing capability.

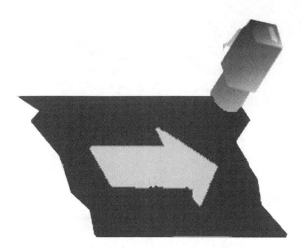

FIGURE 3.4
Image capture using area scan camera.

3.2.3.1 Area Scan Cameras

Area scan cameras are used in a majority of machine vision systems as they can quickly obtain an image of a defined area of an object or scene. The camera has a large matrix of pixels that capture a 2D image of a given scene in one exposure. Figure 3.4 shows a 1-megapixel camera with a zoom lens, which captures the image of the object that appears in its field of view. The distance between the camera and the image determines the field of view. The image quality is best when the distance is adjusted on basis of the specifications of the zooming lens present in the camera. As the resolution of the camera is fixed, it can be easily installed and aligned to capture an image of the desired object.

Area scan cameras are best suited for applications where objects are fixed or stationary, even if only momentarily. To get continuous images, overlapping images have to be captured and cropped using software into individual images and then assembled in the correct sequence.

3.2.3.2 Line Scan Cameras

Line scan cameras use a single row of pixels to build a continuous image. The complete image is constructed line by line as the object moves past the camera. Conversely, the object can be fixed or stationary and the line scan camera can move over the object. Coordination of camera motion and image acquisition timing are critical when using line scan cameras.

Line Scan Camera

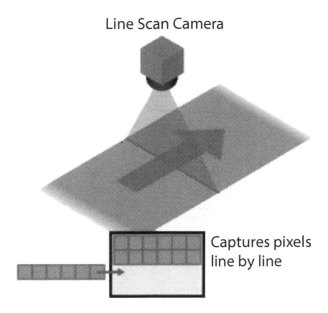

Captures pixels
line by line

FIGURE 3.5
Line scan camera application schematics.

Line scan cameras can be best employed in high-speed processing or fast-moving conveyor line applications. Figure 3.5 shows the use of line scan camera that scans products as they move on a conveyor belt.

Line scan cameras could be used, for instance, to detect defects in continuous materials like cloth or steel chain on the production line. The advantage here is that the inspected material can be processed as a continuous stream. Another classic application where line scan cameras can be used is for inspection of the surface of a rotating cylinder.

Line scan cameras have many advantages when compared to area scan cameras. Line scan cameras use a single row of pixels to build the image, and their resolution is defined by the scan rate. Unlike area scan cameras, the vertical resolution is not limited. Hence, it is possible to have higher resolutions in both 2D and 3D cameras. Also, line scan cameras can be used to capture images of objects that appear between rollers on a conveyor. A single line scan camera can be used instead of multiple area scan cameras to view the opposite side of objects as the camera or objects are rotated. Depending on the application, line scan cameras often require only simple illumination.

Despite these advantages, there are reservations about deploying line scan cameras because of their higher cost. The line scan camera is perceived to be complex because of the precision in timing and coordination that is required for accurate image acquisition.

3.2.3.3 Smart Cameras

Smart cameras are designed for a specific purpose, unlike area scan or line scan cameras which are designed for general use. For instance, smart cameras can be designed for a specific application, such as, to detect motion or to read part numbers or vehicle plates. Hence, smart cameras are being used in many real-world applications including video surveillance, industrial applications, robotics, and games.

Smart cameras use an embedded system for image processing. It should be robust to work reliability in industrial conditions of heat, dust, ambient light, etc. A standard smart vision system is shown in Figure 3.6.

Smart cameras can be easily deployed in machine vision systems as they combine image acquisition and image processing. They would, however, need to be rugged and reliable to work in industrial conditions. As smart cameras are designed for a specific application, it may not always be possible to get smart cameras that suit your requirements. But smart cameras are becoming more popular because they are easy to use and are now being designed to work for many standard applications.

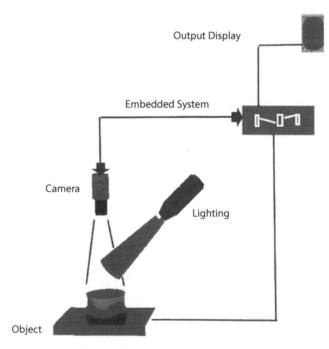

FIGURE 3.6
Standard smart vision system.

3.2.4 Camera Lens

As in the human eye, the lens is used to focus the light reflected from a scene or object onto the image sensors in a camera. The camera then converts the light energy to electrical energy to form the digital image. It is important to understand the relation between the image sensor and the lens, as it impacts the design of machine vision solutions.

A camera lens consists of various parts that are held together in a barrel. At the front is a glass (or plastic) lens that focuses the light. Inside the lens assembly, there are other optical lenses that are used to further refine the image. Each of these lenses is called a lens element. The lens elements are curved because the light reflected from the object is bent and focused on the image sensor. However, the curved glass in the lens can distort the image especially at the edges. Hence, multiple elements are used to counteract this distortion to get a clear and sharp image. Figure 3.7 shows how the interior of a lens is organized.

The lens aperture determines how much light is let through to reach the camera. The aperture opening is adjustable and can be made smaller or larger. Lenses come with fixed aperture settings that are known as F stops. These aperture settings can also be varied. A wide aperture lets in more light and is useful for capturing images of fast-moving objects. However, such lenses are generally more expensive. The lens can also be fitted with a hood, that acts as a visor to regulates the light that falls on the lens. A lens cap is used to protect the lens from dirt and impact. The lens is fitted to the camera using a lens mount. These mounts can be of different types and are discussed later in this chapter.

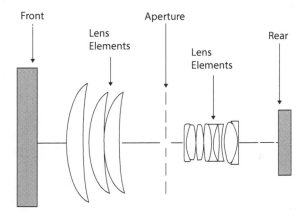

FIGURE 3.7
Skeletal view of a lens interior.

3.2.4.1 Resolution, Contrast, and Sharpness

It is important to match machine vision cameras with appropriate lenses to obtain good-quality images. We know that machine vision applications require quality images with appropriate image resolution, contrast, and sharpness to enable identification of the required features in the image. The choice of lens has a direct impact on the speed and accuracy of image capture. The main challenge is to be able to obtain an image that has clarity and sharp focus throughout the entire image.

We know that resolution is the ability to reproduce the details in the object or scene. The resolution of the image captured by the camera is determined by the number of pixels in the sensor array in the camera. For an area scan camera, the image resolution is marked as N × M matrix, where N denotes the number of rows and M the number of columns in the sensor array. For a line scan camera, there is one row of sensors and hence the resolution is marked as M × 1, where M is the number of sensors in the sensor array. Area sensors are manufactured in different resolutions that typically include array dimensions of 40 × 480, 1024 × 768, 1280 × 1024, 1600 × 1200, 2048 × 1536, and 4008 × 2672. In line scan cameras, M can range from 512 pixels to 12888 pixels or more. Figure 3.8 shows the difference between the images taken under two different resolutions.

It can be observed that high-resolution cameras produce images with greater detail. Sharpness can be defined as the level of clarity of detail in the image. Figure 3.9 shows the sharp and soft images of a captured image.

Viewing distance can be adjusted to improve the sharpness of the image. We learned in Chapter 2 that contrast in an image is the difference in brightness between the light and dark areas. The higher the contrast, the

High Resolution

Low Resolution

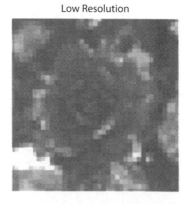

FIGURE 3.8
Images taken under high and low resolution.

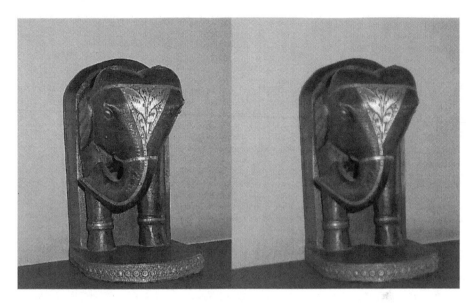

FIGURE 3.9
Sharp image and soft image.

greater is the differentiation between light and dark. Hence, a high-contrast image will appear sharper than a lower-contrast image at the same resolution.

3.3 Lenses and Their Parameters

Several factors have to be considered while selecting a lens for a machine vision application. These include:

- The resolution of the camera sensor
- The distance between the object and the camera
- The size of the object
- The amount of light available for image capture

Resolution of the camera plays an important part in the selection of the lens. For instance, if a high-resolution camera is used for image capture, then the magnification required by the lens is less, as the image can be enlarged digitally. This allows for the selection of a lens with a larger field of view to capture more information in a single frame. Alternatively, if the camera chosen does not have a high-resolution sensor, then you can select a lens with a longer focal length to obtain the detail required for your application.

The distance between the lens front and the object to be imaged is known as the **working distance (WD)**. **Focal length** of the lens is the distance between the lens and the image sensor in the camera. More specifically, it is the distance between the point of convergence of the light rays in the lens to the sensor in the camera. The point of convergence is known as the optical center of the lens, as shown in Figure 3.10.

Focal length of a lens is important because it determines the field of view. The size of object to be imaged is related to the **field of view (FOV)** which determines how much of the scene or object can be imaged. Hence, the bigger the object to be imaged, the larger the FOV should be, and vice versa.

Lenses with a shorter focal length provide a wider field of view of the object but the magnification is less. Conversely, lenses with longer a focal length provide greater magnification but the field of view is less, as illustrated in Figure 3.11.

The focal length of a lens is usually displayed on the lens barrel, along with the size of the adaptor ring. Lenses come with fixed or variable focal lengths. Lenses with fixed focal length are less expensive. Varifocal lenses allow for adjustment to change the focal length of the lens; this adjustment determines the magnification and field of view for a given working distance. FOV and WD share a basic trigonometric relationship dependent on the angular field of view (θ—in degrees) of the system. This relationship can be observed and calculated as shown in Figure 3.12.

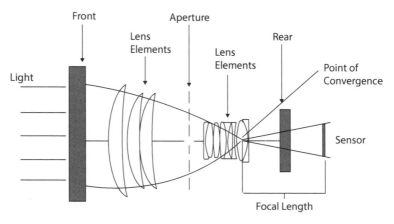

FIGURE 3.10
Lens focal length and point of convergence.

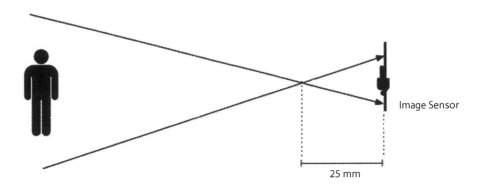

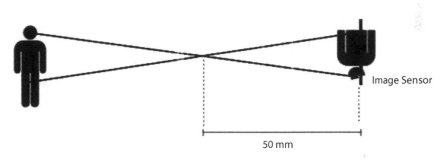

25 mm

50 mm

FIGURE 3.11
Lenses with shorter and longer focal lengths.

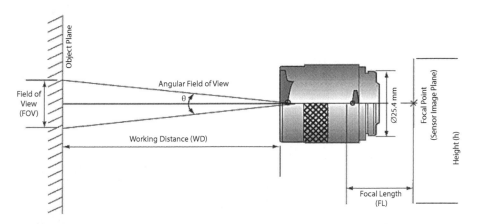

FIGURE 3.12
Relation between FOV and WD.

The trigonometric relationship between FOV and WD is

$$\tan\left(\frac{\theta}{2}\right) = \frac{\text{FOV}}{(2 \times \text{WD})}$$

Similarly, focal length (FL) with respect to height of the object plane (*h*) is

$$FL = h \times \left(\frac{\text{WD}}{\text{FOV}}\right)$$

In machine vision applications, there are often constraints placed on the working distance. For example, application requirements may enforce design limitations on the maximum or minimum distance between the camera and the object to be imaged. In general, a lens that provides a long working distance will be larger and more expensive than one that provides a shorter working distance.

Similarly, the FOV is the object area that is imaged by the lens onto the image sensor. It must cover all features to be measured, with additional tolerance for alignment errors. It is also a good practice to allow some margin (e.g., 10%) for uncertainties in lens magnification. Features within the field of view must appear large enough to be measured. This minimum feature size depends on the application. As an estimate, each feature must have 3 pixels across its width and 3 pixels between features. Hence, if 100 or more features have to be covered, then it would be difficult to do so with a single camera. In such circumstances, multiple cameras may be preferable.

The next aspect to be considered is the **aperture**. Aperture determines how much light is transmitted by the lens to the image sensor. For example, if light is low, a lens with a larger aperture would be preferred as it would allow more light to pass through for image capture. Shutter speed is the length of time the digital sensor inside the camera is exposed to light. A lens with a larger maximum aperture has the same exposure that is obtained with a faster shutter speed. Whereas, a smaller maximum aperture delivers less light intensity and therefore needs a longer or slower shutter speed. A lens with a larger maximum aperture is called a "fast lens"; in comparison, a lens with a smaller maximum aperture is called a "slow lens."

Apertures are described as fractions of the focal length of lens and hence are called f-stops. The f-number is determined by dividing the focal length by the aperture of the lens. For example, a lens with a 50 mm focal length and 10 mm aperture diameter will be a f/5 lens. Figure 3.13 shows lens with different f-numbers.

It can be noticed that as the f-number gets larger, the aperture became smaller. Hence f/2 is bigger than f/4, which, in turn, is bigger than f/8. Therefore, a lens with a low f-number is also known as a better-quality lens,

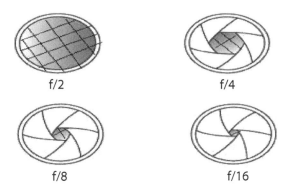

FIGURE 3.13
Lens with different f-numbers.

FIGURE 3.14
Images taken with different lenses at f/4.

though less size is also important. It is also possible to have lenses with f-numbers such as f/2.8, f/5.6, etc.

Figure 3.14 shows an image of a hill view captured with 18, 24, 35, 55, 85, and 105 mm lenses with an aperture size of f/4. From the figure it is evident that the 105 mm lens gives more information compared to the 24 mm lens.

Another parameter to be considered for lenses is the **depth of field** (DoF). It denotes the range of distance between the lens to the object wherein the image remains in focus. In other words, it is the difference between the closest and farthest working distances over which an object may be viewed without any unacceptable blurring of the image. What does this mean? While capturing the image of a scene, say for example a flower, if the background information is also important, then it is necessary to have a wider DoF.

However, if the focus is only on a specific object, like the flower and not the background, then the DoF can be relatively less.

Three factors affect the DoF—aperture, focal length, and the distance between the camera and the image that is to be captured. A wider DoF is possible with a smaller aperture, while a lens with a larger or wider aperture would be more useful when focusing on a specific object in a scene or even a specific part of an object. Focal length is a measure of the lens capability to magnify the image of an object. Hence, DoF would be wider for a lens with a smaller focal length and smaller for a longer focal length. The third factor that affects DoF is the distance between the camera and the object. If this distance is high, then the DoF is wider or larger than when the object is close to the camera.

To sum up, a wide aperture, longer focal length and a short distance between the camera and object to be imaged would provide a shallow DoF. Conversely, a wider DoF can be obtained using a smaller aperture, shorter focal length and moving away from the object to be imaged.

Finally, it is important that the sensor and lens are matched with respect to their size. The image sensor size is typically specified in inches but is not an exact dimension; it simply determines that the sensor lies within a circle of the specified diameter. This is because digital image sensors were originally used to replace the video camera tubes. So a 1″ sensor would fit into a video camera tube with a 1″ yoke. In other words, a 1″ digital sensor means that it is equivalent to a one-inch video camera tube. Sensors comes in various sizes and are also expressed as a fraction of an inch, such as 1/3″, 1/2″, 2/3″. Figure 3.15 shows image sensors in various sizes, wherein, diagonal sizes have been specified. It is necessary to look at the diagonal measurement of the sensor to know the image circle. The diagonal of the sensors is roughly two-thirds (2/3) of the sensor size, as summarized in Table 3.1

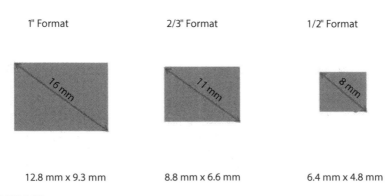

FIGURE 3.15
Image sensors of different sizes.

TABLE 3.1

Sensor Size and Respective Diagonal Size

Size	Width	Height	Diagonal
1 inch	12.8 mm	9.3 mm	16 mm
2/3 inch	8.8 mm	6.6 mm	11 mm
1/2 inch	6.4 mm	4.8 mm	8 mm
1/3 inch	4.8 mm	3.6 mm	6 mm
1/4 inch	3.2 mm	2.4 mm	4 mm

The size of the lens is given by its diameter. This value is found normally on the lens itself, either at the front or at the side near the top of the lens. The size of the lens must be equal to or greater in size to match the circle size of the sensor to obtain the whole image. Figure 3.16 shows different lens sizes matched with a 1/3″ image sensor. If a 1/4″ lens is used, the image circle would not be fully covered. Hence, the image obtained would not be good. A good image is obtained with a 1/3″ lens as the sizes match. It is possible to use a larger lens size such as 1/2″ lens.

To do a recap, selecting a lens appropriate to the camera is essential for obtaining a good-quality image. Lens resolution must fit/match that of the sensor resolution. For a high-resolution camera, the magnification required by lens is less that for a low-resolution camera, which requires a lens with a longer focal length. Focal length also has to be selected based on the working distance available for imaging the required object/scene. Lenses with a shorter focal length provide a wider field of view of the object than lenses with longer focal length, which provide greater magnification. Aperture size

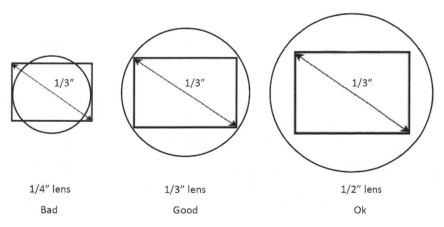

1/4″ lens	1/3″ lens	1/2″ lens
Bad	Good	Ok

FIGURE 3.16
Lens optical format (circle) versus sensor size (rectangle).

is determined based on the available lighting conditions. Aperture must be wider or larger for low-lighting conditions. The DoF is based on the working distance over which the object must be in focus. To get a wide DoF, a lens with a smaller aperture and a shorter focal length is used. Finally, to obtain a complete image, the size of the lens must be equal to or greater in size to match the circle size of the sensor.

3.3.1 Types of Lenses

There are many different types of lenses, including standard, zoom, and telecentric lenses. Each of these can be used for different applications.

Standard lenses, also known as prime lenses, come with a fixed focal length. The focal length is normally expressed in millimeters, such as 50, 85, or 100 mm. The image obtained by standard lenses is comparable to the image as seen by the human eye. These lenses generally have a larger maximum aperture, and hence can work with lower light. Standard lenses are very versatile and commonly used in many machine vision applications.

Zoom lenses have focal lengths that are adjustable over some range. They are useful for prototypes, where the focal length requirement is not yet determined or for applications where a standard lens of required focal length is not available. In contrast to standard lenses, zoom lenses are larger, less robust, more expensive, and have smaller apertures when compared to fixed-focal length lenses. Also, if not used properly zoom lenses can cause distortion in the captured image. Figure 3.17 shows the images taken using standard and zoom lenses.

Wide-angle lenses are used when the frame of the object or scene that is to be captured is large. The field of view that they can capture is wider than that of the standard lens and these lenses have a shorter focal length with a greater DoF. Wide-angle lenses typically

FIGURE 3.17
Image taken using standard and zoom lenses.

FIGURE 3.18
Image taken using standard and wide-angle lenses.

cover a focal length between 24 and 35 mm. Wide-angle lenses are available as primes or zooms and come with either variable or fixed maximum aperture. While they offer a wide field of view, there is the disadvantage of introducing distortion and a shift in perspective of the objects. This is because they can magnify the distance between the foreground and background. Figure 3.18 shows the same image taken using standard lens and wide-angle lens.

Telecentric lenses provide constant magnification for any object distance. They are useful for applications that require high accuracy gauging, such as metrology, CCD-based measurement, or microlithography. Figure 3.19 shows the comparison of images taken using standard and telecentric lenses.

In the first image taken with a standard lens, it can be seen that the distant object looks smaller than the object in front. In the second image taken with a telecentric lens the objects appear the same. Hence, telecentric lenses are used in applications of dimensional measurement or precision gaging and when a constant perspective angle across the field is desirable. The magnification of a telecentric lens is fixed by its design. Telecentric lenses tend to be larger and more expensive than standard lenses.

Fixed Focal Length Lens Telecentric Lens

FIGURE 3.19
Images taken using standard and telecentric lenses.

Superzoom lenses are lenses with very large focal lengths, which cover wide-angle to telecentric lenses. They can be used in situations with a large degree of variability and high mobility.

Macro lenses are camera lenses optimized to work at magnifications near 1X. A 1:1 magnification means that the size of the image is the same size as reproduced on the sensor. Macro lenses are available in a wide range of focal lengths. The longer the focal length, the greater the working distance. When taken at short working distances, the image sharpness would be high, but the depth of field would be tiny.

For magnifications greater than 1X, camera lens can be used in reverse, with the object held at the usual camera plane and the camera in the usual object plane. In this case, the object distance will be short, while the lens-to-camera distance is long. Adaptors are available to hold camera lenses in this orientation.

3.3.2 Lens Mounts

Lens mounts are used to attach the lenses to cameras. These mounts provide mechanical stability and define the distance between the sensors and image. Lenses can be proprietary, which means that certain lenses can be used only with cameras made by the same manufacturer. Therefore, to ensure compatibility between cameras and lenses, the appropriate lens mount must be used.

Presently, mounts generally come in three major types: C-mount, CS-mount, S-mount, and F-mount. The sizes for each mount are based on industry standards.

The mount size determines the shortest possible distance between the lens and the image sensor. It is necessary to know which mount your camera uses. For example, it is possible to use a C-mount lens with a CS-mount with an extension tube. However, a C-mount camera cannot be used with a CS-mount lens.

The C-mount is the most common type of lens mount used in machine vision applications. It can be used with a range of lenses and accessories. For instance, the back focal distance of a C-mount connector is 17.526 mm, while that of a CS-mount is 12.526 mm. Hence, a C-mount lens can be used on a CS-mount camera with an extension tube or adapter of 5 mm fitted between the camera and lens.

C-mounts and CS-mounts have a small diameter, so they do not work well with high-resolution cameras or line scan cameras. The F-mount is commonly used for line scan cameras. There is also a T-mount, also known as M42 mount, that is used as an alternative to the F-mount for line scan cameras and other high-resolution cameras. These mounts are considered to be robust.

C-Mount   CS-Mount   F-Mount   S-Mount

FIGURE 3.20
Camera mounts.

The S-mount is a standard small lens mount used in various surveillance CCTV cameras and webcams. These lenses allow only a small amount of adjustment. It uses a male metric M12 thread with 0.5 mm pitch on the lens and a corresponding female thread on the lens mount; thus an S-mount lens is sometimes called an "M12 lens." This kind of lens mount is also used in microbiology research. Different lens mounts are shown in Figure 3.20.

3.3.3 Lens Selection Examples

Let us look at examples of how to choose a lens for a camera. For this purpose, we classify machine vision lenses into two broad categories:

- Lens for field of view (image size) that is much larger than camera sensor size
- Lens for field of view that is smaller or near to camera sensor size

To choose a lens, we have to consider the three important lens parameters in the selection process:

1. Field of View
2. Working distance
3. Sensor size of the camera

We can derive the following formulae from the above parameters:

$$\text{Magnification of a image acquired} = \frac{\text{Sensor size of the camera}}{\text{Field of view}}$$

$$\text{Focal length} = \text{Magnification} \times \frac{\text{Working distance}}{1 + \text{Magnification}}$$

3.3.3.1 Field of View (Image Size) Is Much Larger Than Camera Sensor Size

For applications in this scenario, the field of view (FOV) can typically range from about 20–30 millimeters to as large as 100 meters for outdoor applications. Let us assume that we have a camera with a sensor size of 2/3″. If an application requires the camera to look at a FOV of 50 mm (horizontal) and a working distance of about 200 mm, by using the equations mentioned earlier we should be able to estimate a lens suitable for this application as follows:

FOV = 50 mm

Sensor size = 8.8 mm (based on 2/3″ sensor size)

Working distance = 200 mm

Magnification = 0.176

Estimated focal length = 29.93 mm

As lenses come with certain fixed focal distance, it would not be possible to find a lens with the exact focal length. Three options are available for the appropriate lens.

1. Select a 25 mm lens to provide your application with a bigger field of view at the expense of resolution.
2. Increase your working distance by selecting a 35 mm lens (moving your camera and lens farther away from the object you are looking at increased FOV).
3. Select a 25 mm lens and add an extension ring to it. The extension ring to be chosen to provide the additional focal length of 4.93 mm (29.93–25 mm).

3.3.3.2 Field of View Is Smaller or Close to Camera Sensor Size

Telecentric lenses come with magnification from about 0.01× to several hundred times that of the image sensor. Most such applications use telecentric lenses and these lenses provide low distortion and near exact image reproduction.

However, telecentric lenses are seldom or never used in large FOV applications. This is because in order to get low distortion, the physical lens needs to be as big as the image you are looking at. These lenses are specified in terms of magnification (image size/sensor size) and working distance is usually rigidly fixed by the lens maker.

To determine an appropriate lens, calculate the appropriate magnification and select the nearest magnification that matches your requirement.

Sample Calculation:

Let sensor size = 290 mm

FOV = 28.7 mm

Distance from sensor to measuring region (WD) = 605 mm

Then magnification factor β = FOV/Sensor length = 28.7/290 = 1/10.1 = 1/10 (approximately).

With magnification β and the distance from sensor to measuring region D, focal length of a lens can be derived using the following formula:

$$\text{Focal length } f = D/(1/\beta + 2) = 605 \text{ mm}/(10.1 + 2) = 50 \text{ mm}$$

Hence, we have to choose 50 mm lens for best results.

3.4 Machine Vision Lighting

In this section, we shall look at different illumination techniques that help to yield a good-quality image with a high degree of repeatability. We know that images are created when light from an object is reflected and captured by a camera as shown in Figure 3.21.

The part of the object that reflects light toward the camera appears bright. In other words, if the light is reflected away from the camera or is absorbed

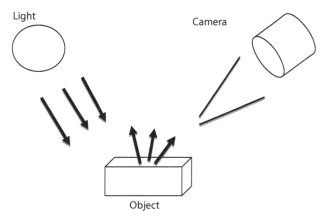

FIGURE 3.21
Setup for image capturing with lighting.

by the object, then the respective part of the object appears dark. The lighting must therefore be uniform and consistent and controlled to provide a clear image that is not too bright or too dark.

Lighting is chosen based on the size, shape, color, and texture of the object to be imaged. When we talk about illuminating an object, we refer to two distinct aspects: the light source and the lighting technique. For instance, lighting can be used to silhouette a part of an image to allow measurement of its edges. A diffuser can be used to spread light over an object, while a collimator is used to make the light rays parallel. A polarizer is used to reduce glare reflected from the surfaces of objects.

3.4.1 Light Sources in Machine Vision

Different types of light sources are available for use in machine vision applications. Lights in machine vision applications can be constant or strobed. In strobed or pulse mode, lighting is switched on only when required. Strobe light is therefore short intense flashes of light whose length and intensity can be configured. Strobing can increase the life of the light source, but the duration of the strobe has to be carefully controlled and timed with image acquisition. We will now look at some of the commonly used light sources.

LEDs (light-emitting diodes) are currently used extensively in machine vision applications. The output energy of a single LED is low, so they are combined in arrays. The light from an LED array can be directed or focused as required. LED lights are very safe and efficient and have a long life span. They are available in different colors—visible, ultraviolet, as well as infrared. These lights are vulnerable to damage due to the heat they produce. It is also difficult to use LED lighting to illuminate large areas.

Fluorescent lamps (we know tube lights) produce light by a chemical reaction. Electric current is passed through the mercury gas filled lamps/tubes to produce ultraviolet light that reacts with the phosphor coating inside the glass to emit fluorescent or glowing white light. A ballast is used to provide a high initial voltage to start the process and then limit the lamp current to sustain the light. For machine vision applications, a high-speed ballast is used to drive the lamps to reduce flicker and make the light intensity consistent. These lamps are relatively inexpensive, are available in many shapes and sizes, and have a reasonable life span. This light is best suited for diffuse lighting, that is, covering a larger area around the object to be imaged. Unlike LED arrays, they not suitable for directional lighting.

Another type of light source is halogen lights. They are useful for applications where bright light is required. In halogen lamps a live conductor like tungsten is heated in a protective gas atmosphere to produce electromagnetic light. Halogen lamps can be used where large areas or lots of light is required for industrial image processing. A halogen light can be combined with a reflector to form a point source. When used in a machine vision application, as halogen lights tend to be hot, illumination can be delivered using optical fibers. The lamp can be mounted remotely to provide bright and uniform light required for line scan camera applications.

Lasers (Light Amplification by Stimulated Emission of Radiation) provide high-intensity light sources. A laser is created when electrons in atoms are excited using electric current. The electrons absorb the energy and when returning to their original state give out photons of lights through a process known as spontaneous emission. Lasers are very powerful but can be easily focused and controlled. They are widely used to provide structured light, such as a dot, circle, line, or any other geometric pattern. Laser lights are widely used for precision measurements. They are expensive and require safety precautions as they are hazardous to health.

Many of the older systems used xenon light for strobing systems. It is capable of giving high-intensity light over a short period of time. Metal-halide light is produced by an electric arc in a gaseous mixture of mercury and compounds of metals like bromide or iodine.

To summarize, fluorescent and quartz halogen lighting sources were originally used in many machine vision applications. Current machine vision applications widely use LED lights, which are cost-effective and efficient. Laser lights are very effective, but safety standards have to be stringent.

An important consideration when choosing the light source is the ambient or background light present in the room. Common sources of ambient light are factory lights, sunlight, and other specific lighting at inspection points. Ambient light can vary depending on time of day or weather or season. The light used to illuminate the part that needs to be inspected should be bright enough to offset the ambient light to obtain a good-quality image and maintain the consistency of inspections.

3.4.2 Illumination Techniques

Lighting technique refers to the placement of the light source with respect to the object/part to be inspected and the camera. Different types of lighting techniques can be used depending on the requirements of the application and the presence of ambient light.

3.4.2.1 BackLighting

Figure 3.22 illustrates backlighting. The camera is at the front of the object. The diffuser and light source are placed behind the object.

Backlighting can be used to create a silhouette of the object. The shape of the object is outlined but surface details may not be clear. It also produces an image with high contrast. Sample images taken using backlighting are shown in Figure 3.23.

As evident in Figure 3.23, the contrast created by backlighting enables the letters etched on the bottle to be read. In the second figure, light penetrates

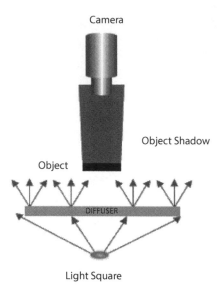

FIGURE 3.22
Backlighting.

FIGURE 3.23
Images taken under backlighting.

through the holes in the object enabling the holes to be measured. Fill levels in bottles can also be measured, as there is difference in the light reflected by clear bottle and the liquid-filled part.

3.4.2.2 FrontLighting

In frontlighting the light source and camera are at the top of the object as shown in Figure 3.24.

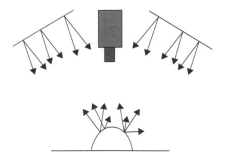

FIGURE 3.24
Structure for frontlighting.

FIGURE 3.25
Image taken under frontlighting.

This type of lighting is easy to set up and provides good contrast. However, shadows may be created which can be avoided using multiple lights. A ring light can be used, for instance, to direct light onto the object. Figure 3.25 shows an image taken under frontlighting.

Directional frontlighting can be either bright field or dark field lighting. In directional field illumination, the incident light is reflected by the surface of the target. The angle of incident is generally between 30 and 60 degrees with respect to the object surface. Figure 3.26 shows the dark and bright fields. Bright field is the region where reflected light is within the field of view (FOV) of the camera. Dark field is the region in which reflected light is outside the FOV.

An image taken under directional field illumination is shown in Figure 3.27. It can be seen in the figure that directional field enhances the edges, which is required for recognition of objects.

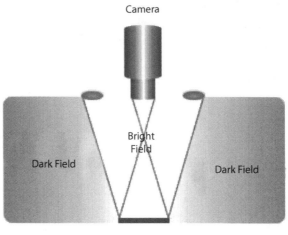

FIGURE 3.26
Directional field illumination.

FIGURE 3.27
Images taken under directional field illumination.

3.4.2.3 Diffused Lighting

Diffused light eliminates reflections and shadows, as light is directed onto the object from multiple angles. Diffusers provide a large illumination area that avoids glares. Diffusers can be used with the other types of lighting explained above like backlighting and frontlighting.

Diffused dome lighting is shown in Figure 3.28. The dome is used to provide evenly distributed uniform illumination to avoid shadows.

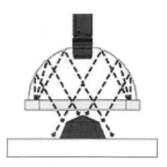

FIGURE 3.28
Diffused dome lighting.

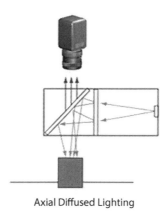

Axial Diffused Lighting

FIGURE 3.29
Axial diffused lighting.

In axial diffused lighting, the camera is above the object to be imaged. The light source illuminates the object from the side, as shown in Figure 3.29.

A semitransparent mirror (50% silvered mirror) is used to cast the light downwards on the part. The camera can image the object through the mirror. This type of lighting can be used to detect flaws on surfaces and to inspect the insides of cavities.

3.4.2.4 Oblique Lighting

In oblique lighting, the light falls with different angle of incidence from 0 to 15 degrees with respect to the object surface. This lighting helps to identify textural details such as dust or edges. Sample images are shown in Figure 3.30.

FIGURE 3.30
Images taken under oblique lighting.

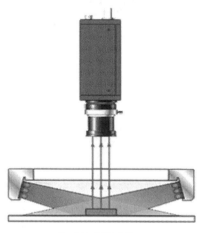

Dark Field Lighting

FIGURE 3.31
Dark field lighting.

3.4.2.5 Dark Field Lighting

Dark field lighting, shown in Figure 3.31, is used to create a bright, easily detectable feature of interest within a dark background.

For example, consider light rays that are directed at a transparent bottle from a dark field region. Most of the rays pass through the transparent object undetected by the camera. If there are irregularities, such as a crack in the bottle, light gets scattered in many directions, some of which are reflected to the camera. Dark field lighting can be used for detecting surface defects like scratches, burrs, or other small raised features. A sample image taken under dark field lighting is shown in Figure 3.32.

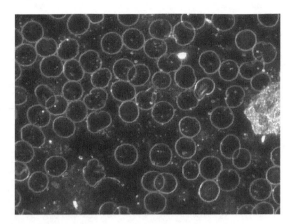

FIGURE 3.32
Image taken under dark field lighting.

3.4.2.6 Infrared and Ultraviolet Light

Light beyond the visible spectrum can be used in machine vision applications. Infrared light can be used to moderate the grayscale difference in colored objects. Dark objects absorb infrared light waves to obtain uniformity when there are varying shades. Ultraviolet (UV) rays can be used for imaging small features in an object. For example, bruises on fruit surfaces could be detected using UV light.

3.4.3 Illumination Summary

We have looked at different light sources and lighting techniques. We want to control the lighting environment in order to obtain a good-quality image that contains and emphasizes the information needed in the application. Accurate and consistent lighting will enable standardization in image capture and therefore repeatability of the inspection results.

One of the important considerations when choosing illumination is the size of the object/part that is being imaged. If the part is small, say for example a 1-inch square, LED or fiber optic lighting may be best. However, if area to be illuminated is bigger, then fluorescent lighting may be chosen.

If the part is static (not moving), then a constant lighting may be used. For moving parts, a strobe light source would be preferable as it helps to freeze the motion.

The shape of the part can also influence the type of lighting. Ring lights are best used to illuminate circular or rounded objects. Square or rectangular parts are best illuminated using line lights. The surface texture of the part also determines the type of illumination. For shiny or glossy surfaces, diffused lighting is preferable as it eliminates surface glare.

We discussed different types of lighting and their suitability for various applications. Oblique or dark field lighting is best suited for looking at surface defects on parts/objects. For example, if there is crack in a glass surface, cracks would show up as bright feature on a dark background. To detect missing, broken, or incorrect objects with respect to shape or color, bright light can be used. Dark lighting is suited to inspecting transparent packaging as this type of lighting eliminates reflection and provides a high-contrast view of the object. For dimensional measurements, backlighting is used as it accentuates the shape of a part, while obscuring the surface detail.

The following aspects need to be looked at with reference to light sources and techniques:

- Light output can deteriorate over time and the lights need to be replaced regularly.
- Lights produce heat when used. This can affect the consistency of inspection.
- Variations in power supply to the light/lamp can affect the light consistency.

3.5 Filters

Image formation captured through cameras can be optimized with the help of optic filters. which can be used to enhance the contrast in the resulting image. This can, for instance, be achieved by suppressing interfering wavelength ranges in the light spectrum in order that colored features can be lightened up or darkened when using colored lighting. Interfering effects of inappropriate light can be avoided, or the wavelengths of the light used can simply be limited to the extent that the camera sensor works at its optimum. Filters can be of different types:

- Color filters to emphasize or suppress specific wavelength ranges and colors
- Grey filters as "sun shades" to reduce the overall brightness
- Interference filters as band-pass filters for specific wavelength ranges
- Polarizing filters to avoid reflections on test objects
- Heat filters as protection against infrared and heat radiation
- Light control film to block diffuse background light

Except for light control films, filters are usually mounted to the filter thread of the lens. If this is not available, special plug-on mounts are available. One example of this is telecentric lenses. In some industrial cameras, filters can be mounted between the lens and the sensor. The advantage is that the filter can then be rather small and thus cheaper than a very large filter in front of an objective with large front lens. However, this is possible only if option is provided by the camera manufacturer.

Color filters are made of colored glass that absorb light in specific wavelength ranges in varying degrees and let other ranges pass to a major extent. Depending on the technical function and translucence in specific wavelength ranges, these absorption filters are called color filters, daylight elimination filters, high-pass filters, etc.

Color filters are typically used for monochrome cameras to filter light from the entire spectrum in such a way as to produce something like colored illumination. The filter glass can be red, green, orange, blue, or yellow. Wavelengths of their own colors are allowed to pass to a large extent (say 90%) in comparison to other colors that are blocked. A portion of the rays are also reflected at the glass surface. If a color filter is used with white illumination, there can be a significant loss of light. Filters work according to the principle of subtractive color mixing, and due to their design are not as narrow-banded as interference filters.

Using white light as a basis, filters can be used to filter required colors. In this way equally bright green or red, which could not be distinguished in the monochrome camera image, can be differentiated; by using a red filter, the red can be lightened up and the green of the test object can be darkened.

3.6 Machine Vision Software

In the previous sections, we learnt about different types of hardware components that are used to build a machine vision application. Another equally important component of machine vision applications is software. Software drives image acquisition, processing, and analysis functions and is responsible for the overall performance of machine vision applications. Choosing the right software can make the difference between a wasted investment and a highly effective application.

Software needs to be assessed not only in terms of technical capabilities but also with respect to speed of processing and other performance parameters such as ease of use, maintainability, reliability, etc. For instance, a machine vision system for a high-speed application that requires excellent image quality will generate high volume of image data. Hence, it would be necessary to choose machine vision software with good processing power and memory capabilities. For a standard application, off-the-shelf software, an embedded vision system with built-in software, may be the best route, as integration, implementation, and ongoing management are all relatively simple.

Some of the other important considerations when choosing machine vision software are discussed below.

3.6.1 Integration and Compatibility

In today's manufacturing environment, nothing is totally isolated, and this is especially true of machine vision systems. Whether the machine vision software needs to save images and data, communicate with a remote interface, or control actuators to sort products, its ability to integrate with other components is critical. This is true for hardware components as well.

3.6.2 Ease of Use and Cost to Operate

While considering cost of development, the purchase price of the software would be routinely budgeted. An interactive development environment (IDE) is provided in most instances for developing the application. More often than not, certain routines or modules have to be specially written to meet some of the required functionalities of the application. Regular updates and maintenance of these programming modules would add to the time and cost of operating the application.

3.6.3 Vendor Support and Stability

It is important to carefully choose the software vendor. Even if the software is easy to use, assistance will be required for clarifying doubts and for answering queries regarding the use of the various software options provided. A company with knowledgeable and responsive technical support can provide timely and proper assistance and help to avoid costly downtime. Additionally, it is necessary to establish the credentials of the vendor as well their market share. A company with a sizable market share is likely to be stable and remain operational.

3.7 Machine Vision Automation

Industrial automation is vital to meet the ever-increasing demand for productivity, better quality standards, better accuracy, and optimum utilization of available resources and manpower. The automation system must be flexible, cost-effective, and efficient.

In machine vision applications, we saw that there are three main steps. First is image acquisition, next is image analysis or processing, and the third

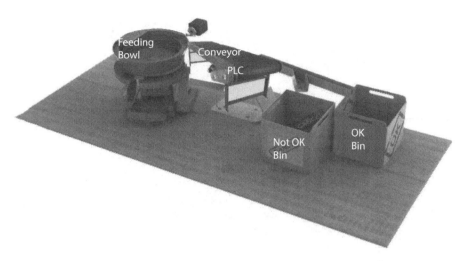

FIGURE 3.33
Simple automation setup.

step is acceptance or rejection based on the output of the analysis. In this section, we look at the third step, and the components generally used to carry out this process. Industrial automation replaces human decision-making with the use of mechanized equipment and logical instructions/commands. Figure 3.33 shows a simple automation setup consisting of conveyer, PLC, and the OK and not OK bins.

The conveyor belt is used for transporting the parts/objects that need to be inspected. After the inspection process, as the objects continue to move on the conveyer, they undergo scanning by the sensors which sends the information to PLC (programming logic circuit). Based on the image analysis results, the PLC automates the segregating conveyors to direct the products into the respective bins.

The rotating mechanism seen in the figure, allows the conveyor belt to transport the products to various stages. The conveyors are connected to the rotating mechanism in various directions to transport and direct the products efficiently. In industrial settings, conveyors of different length, sizes, and shapes are used to transport materials that match the product/part specifications as well as the inspection requirements.

In automation, sensors play a major role in segregating the products based on the respective parameters to various stages. The sensors used can be of any type depending on the type of applications and their requirements. Sensors help in finding the midpoint by using the high-level and the low-level signals and communicate the findings to PLC.

Sensors can be used to sense the part and inform the PLC which can trigger the camera(s) to capture the image. Likewise, sensors can be used to sense the part as it comes out of the image processing section and inform the PLC. The PLC can activate the output switches based on the image processing results to segregate the good from defective parts.

The PLC is a digital computer that is used for automation for the electro-mechanical parts such as the conveyor, assembly lines, and robotic devices. They are specially built to be rugged for use under harsh conditions of heat/cold, pollution, and moisture common in industrial environments. Figure 3.34 is a block diagram of a PLC.

The program that controls the automation is written separately on a standard computer and is then downloaded and stored in the memory of the PLC. The input and output modules connect the PLC to the outside machinery. The PLC receives information from the input devices, processes the data in the CPU, and triggers the output based on the programming outcomes. The input devices can be sensors, switches, or meters, while the output devices can be actuators, lights, valves, etc. For example, a PLC can monitor recorded temperatures to automatically start or stop processes and generate alarms if machines malfunction.

The PLC offers a range of ports as well as support for protocols to be able to communicate with other systems. For example, application data in the PLC may be required to be transported to an external computer for storage and further analysis. The PLC may also contain an HMI (human-machine interface) with a display unit/touch screen and a simple keypad to view and input information needed by the PLC in real time.

Quality control is one of the primary areas where machine vision is extensively used. Other areas of industrial automation include automation of manufacturing and material handling processes. Automation is often customized to meet the needs of a particular application. For example, if the application is inventory management, the automation component could be a robot with interface for label inspection and barcode reading. The robot could be equipped with a mechanical arm to handle stocks for

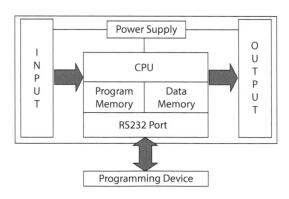

FIGURE 3.34
Block diagram of a PLC.

sorting and storing. Robot guidance systems use vision to guide the robot on wires in the floor. In factories, Automated Guided Vehicles (AGV) are commonly used to transport goods and material. AGVs are programmed to follow routes or paths to locations where materials are to be picked up or deposited.

When designing an automation system, it is essential to calculate the overall time that will be needed to complete the process. The entire system must be designed with precision to enable parts to be imaged and analyzed. The outcomes of the analysis must be available to allow for decision to be executed in real time.

3.8 Integration of Machine Vision Components

Since vision systems often use a variety of off-the-shelf components, these items must coordinate and connect to other machine elements quickly and easily. Typically, this is done by either discrete I/O signal or data sent over a serial connection to a device that is logging information or using it. Discrete I/O points may be connected to a programmable logic controller (PLC), which will use that information to control a work cell or an indicator such as a stack light or directly to a solenoid, which might be used to trigger a reject mechanism. Data communication by a serial connection can be in the form of a conventional RS-232 serial output or Ethernet.

Some systems employ a higher-level industrial protocol like Ethernet/IP, which may be connected to a device like a monitor or other operator interface to provide an operator interface specific to the application for convenient process monitoring and control.

3.9 Summary

In this chapter, we looked at a typical machine vision system and the various components that are used to build an application. We learned about different types of cameras (image sensors) that are used to capture images of parts/objects. The camera is chosen to match the needs of the application. We saw, for example, that area scan cameras can be used effectively for scanning objects that are fixed or momentarily stationery. Line scan cameras are best employed for high-speed processing where the inspected material is processed as a continuous stream.

We saw the importance of matching the camera and lens to obtain good-quality images with appropriate image resolution, contrast, and sharpness.

We learned about the various lens parameters like focal length, working distance, aperture size, and depth of field and the relationship between these parameters. We looked at different types of lens and their application to capture different types of images.

Proper illumination is another critical aspect of good-quality image capture. Different sources of lights and lighting techniques can be used in placement of the light source with respect to the object/part to be inspected and the camera. Illumination choices will vary depending on the size of the image, whether the images are static or moving, and the type of object surface such as glossy or matte. We also saw how filters can be used to suppress certain wavelengths in the light spectrum to darken or lighten features.

Machine vision components are available in a wide variety of off-the-shelf components. It is important that these components are properly connected and integrated using communication channels to meet the specific application needs. Last, but not the least, we saw the importance of choosing the right software for image analysis. Appropriate software enhances the overall performance of the system while providing the right functionality and ease of use.

Automation is the next step in inspection process that takes the results of image analysis and uses for decision-making and/or controlling the process. The main areas where machine vision is extensively applied in industrial automation is for quality inspection/control, automation of the production processes, and in stock management. Automation is a vast topic by itself and we have only attempted to provide an introduction to automation to understand how it fits in the overall machine vision system. More often than not, automation is customized to meet the application needs.

Exercises

1. What are analog cameras? What are the differences in the working of analog and digital cameras?

2. Explain CCD and CMOS image sensors using their architectural layouts. What are their advantages and drawbacks?

3. Explain progressive and interlaced scanning with reference to area scan cameras. What are the advantages and disadvantages of the two modes of scanning?

4. Explain the differences in the working of area scan and line scan cameras.

5. Discuss two examples of the application of both area scan and line scan cameras. Explain the rationale for choice of camera.

6. What are the benefits of using a smart camera? Explain with examples.

7. What is a telecentric lens? Draw a sketch of the basic light rays for a telecentric lens and explain its function. Where is it used?

8. Explain the anatomy of a lens interior with a neat and simple diagram.

9. What factors should be considered in selecting a lens for a machine vision application? Derive the relation between FOV, WD, and focal length.

10. Consider the scenario where the FOV can typically range from about 50–70 millimeters to as large as 100 meters for outdoor applications. Let us assume that we have a camera with a sensor size of 1/3″. If an application requires the camera to look at a FOV of 60 mm (horizontal) and a working distance of about 250 mm, estimate a lens suitable for this application.

11. Consider an outdoor application where the FOV can typically range from about 10–20 millimeters to as large as 50 meters. Let us assume that we have a camera with a sensor size of 2/3″. If an application requires the camera to look at a FOV of 50 mm (vertical) and a working distance of about 250 mm, estimate a lens suitable for this application.

12. "Proper illumination is a critical aspect to good quality image capture." Justify the statement.

13. What is the difference between front- and backlighting? Where would you use the two types of lighting? Explain with examples.

14. Identify the illumination techniques for the following diagrams.

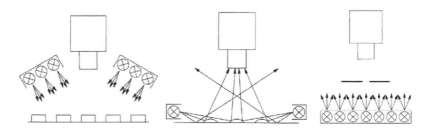

15. What factors should you consider when choosing the light source and lighting techniques?

16. What are filters? How can they be used to improve an image?

17. What are polarizing filters? Explain their usage with examples.

18. What are color filters? What colors are transmitted if you use the following filters on white light?

    a. Red filter

    b. Blue filter

    c. Red and green filter

19. Draw the block diagram of a PLC and explain its workings. Explain the use of a PLC with a sample application.

20. List the factors to be considered when selecting machine vision software.

4

Machine Vision Applications in Quality Control

In the first three chapters, we have seen how machine vision works and discussed the components that are used to build an application. We have also looked at digital images, their characteristics, and some basic operations on digital images. In this chapter, we will look at some actual applications of machine vision in quality control. While we assume that the readers are familiar with quality control and its principles, we start with a brief overview to put the ensuing discussions in proper perspective.

Machine vision is used in the industry for quality inspection, automated defect detection, and segregation of good and defective products. In this chapter, we discuss the application of machine vision for quality control for different industries, such as, automobile industry, food industry, and drug and pharma industry. We will also discuss quality inspections in the industry using examples.

4.1 Overview of Quality Control

Quality control has always been inherent to manufacturing and customer satisfaction. The prime objective of manufacturing is to make products that customers can buy and use. Hence, it becomes necessary to manufacture the products that the customers need at the right quality and at the right price.

Before, we look at a formal definition of quality control, let us take a brief look at how quality control has evolved over the years. In the early days, quality control was restricted to the people who made a product. They dealt directly with their customers and were responsible for the quality of their products. The Industrial Revolution ushered in the era of mechanization of the manufacturing process. Quality control departments were set up to inspect the products for defects and completeness. Finished products had to pass quality checks prior to their dispatch to customers. Inspection was primarily manual, though quality inspectors used tools and techniques to test the assemblies, subassemblies, and finished products. Quality checking was usually not 100%; instead, random samples (on scientific and statistical basis) were tested for quality.

As manufacturing processes and techniques evolved, a more general view of quality was proposed. Quality control that focused solely on customers and their requirements was extended to include more interested parties, both within and outside the company. Other parties involved in the quality process included the employees, the suppliers, the investors, the transporters of goods, and many others. As a result, the concept of quality control evolved from being exclusively the responsibility of inspectors to that of the company. This concept was defined as Total Quality Control (TQM). TQM in the form of statistical quality control was invented by Walter Shewhart and was championed by W. Edward Deming, who is considered to be the father of TQM. TQM was further developed in Japan in the 1940s by Deming, Joseph Juran, and A.V. Feigenbaum.

TQM can be understood as a framework that promotes quality in an organization. TQM creates a permanent climate in an organization to deliver high-quality products that meet the expectations of customers. It also includes its ability to continuously improve the quality process using feedback obtained from customers, suppliers, and other interested parties. The International Organization for Standardization (ISO) defines the requirements and guidelines for quality management systems in its ISO 9000 standard. Use of this standard helps organizations to demonstrate their ability to provide products and services that meet customer requirements with continual improvement. There are many other well-known methodologies, like Six Sigma and lean manufacturing, that provide organizations with tools and practices to establish, manage, and continuously improve their quality management system.

Quality has therefore emerged as a continuous process, starting from conception to the final product through the whole manufacturing process. Quality control (QC) focuses on identifying defects in products and monitoring activities to verify that products meet quality standards. Quality assurance (QA) aims for defect prevention by focusing on the process, the techniques, and methods used to conceptualize and design the products. QA/QC is the combination that ensures products and services are designed and manufactured to meet the customer expectations.

4.2 Quality Inspection and Machine Vision

Quality inspection covers the full cycle of manufacture—raw material, components, and parts at various stages of production to the final product. Inspections help to identify defects early in the process, thereby reducing manufacturing costs. Inspection can be carried out to compare available raw materials, components, or products with established manufacturing specifications. If an inspected item falls within the specified or allowable range, then the item is accepted, otherwise it is rejected. Quality control helps to improve product quality, automate production processes, and to reduce the cost of manufacture.

Machine vision (MV) technology uses digital cameras and image processing to replace or complement manual inspections and measurements. Quality inspection of components is labor intensive and can be prone to error. Also, 100% manual inspection of manufactured parts/products would be expensive and time-consuming and may not be feasible. Machine vision provides a fast, economic, reliable, and objective inspection technique. It is also an automated, nondestructive, and cost-effective technique that can be used across industries for many different applications. Some of the key benefits of using machine vision can be enumerated as follows:

- **Accuracy:** Machine vision systems provide greater precision and accuracy of quality inspection when compared to human operators. Even when humans rely on a magnifying glass or microscope, machines are still more accurate because they can "see" and measure parts to a higher tolerance.

- **Speed:** Parts can be inspected at higher speed compared to inspection by human operators. In addition, such high-speed inspections can be carried out with greater efficiency and productivity.

- **Repeatability:** Quality inspections can be repeated in the exact same way by machine vision systems without fatigue, 24/7 and 365 days in the year. In contrast, human inspectors may arrive at different measures at different times, even if all the parts are exactly the same.

- **100% Inspection:** Machine vision systems can be designed to inspect 100% of the parts or products manufactured. Otherwise it might only be possible to inspects parts/products that are randomly sampled using statistical methods and tools.

- **Cost:** Because machine vision systems are faster than humans, such automated inspections can reduce costs. The accuracy of the process helps to reduce the dispatch of defective parts to customers. This in turn reduces external failure costs such as replacements, loss of sale, and warranties.

One of the main advantages of a machine vision system is its noncontact inspection methodology. This is particularly useful in cases where it is difficult to implement contact measurements. Machine vision systems can also be applied in environments that are hazardous to human inspection.

4.3 Designing a Machine Vision System

Let us recall how a machine vision system works. The vision system captures an image of a part or product and sends the image to an image processing unit, such as a PC, where the image is analyzed. This analysis could be measurement of dimensions, reading or verifying a label or code, or identifying presence of defects in the part or product. The results of image analysis are used to control factory equipment such as programmable logic controllers (PLCs) or robots. The analysis has to be carried out quickly so that action can be taken in real time; for example, segregation of parts into respective bins based on color, shape, size, or type of defects.

Therefore, the first step in designing a machine vision application is to determine precisely what it needs to achieve. For instance, some of the key questions that need to be addressed include:

- What parts or products have to be inspected?
- Are the inspections continuous (100%) or discrete (sampling)?
- What should be the speed of inspection?
- What are the performance requirements?

For example, for 100% inspection, the speed of inspection must match the speed at which parts are produced and pushed onto the conveyer belt for inspection. If parts are randomly sampled, then it is necessary to arrive at the number or percentage of parts to be inspected and how frequently the parts have to be inspected. Quality application may require an uptime of 95%—99% depending on the industry and application. Frequent breakdown of quality inspection would lead to production delay and consequently the dispatch of goods may be delayed.

It is essential, therefore, to understand the characteristics of the products or parts that are to be inspected. For instance, if the dimension of a manufactured part is measured for correctness, then it is necessary to know the level of tolerance or variability that can be present for each dimension. For example, only a tolerance level of 5% may be allowed, as otherwise the part may not fit correctly in the component/assembly. Likewise, in the food industry, a fruit, such as an apple or mango, may be inspected for blemishes to determine quality. Such blemishes, for instance, may indicate the presence of worms or bacteria and such fruit may not be chosen for export or local sales.

Having a proper inspection solution in place provides confidence that consistent product quality levels are met during the manufacturing process as well as at final product assembly. Some of the major advantages of this kind of inspection include:

- Reduction in end-line defects
- Saving in time and effort in final inspection

- Problems can be fixed at the beginning with resulting saving in cost of rework
- Prevention of common mistakes or mistakes that are often repeated

Machine vision system components and configuration have to be carefully designed to match the needs of the quality application. We studied the different types of components in Chapter 3, and we know that each of these can be used for different areas and applications. For example, the quality of material inspected can determine the type of camera and lighting that needs to be used, as each of these materials—glass, paper, plastic—have different light reflective properties. Likewise, line scan cameras may be preferred over area scan cameras for measurement of continuous material. In some cases, the applications may require complex lighting or custom design of material handling logistics.

We know that 3D images, in comparison to 2D images, may contain a higher level of topographical, chromatic, and textural information about objects. However, not all applications require this high level of detail. For example, to check for the presence or a burr in a product, a simple 2D greyscale image may be enough. On the other hand, a 3D color image may be required for certain health-related applications, such as robot-controlled operations.

Machine vision systems can, therefore, yield significant benefits in the quest for improved quality, including reduced scrap and rework, improved productivity, greater product reliability, increased consumer safety, and greater customer satisfaction. However, a word of caution is warranted. While all physical objects can be identified by capturing images, it may not be possible currently to measure other nonvisual and physical characteristics of the object such as the mass, sound, or vibration.

Machine vision systems have been in use for quality control for more than a decade. The earlier systems were very expensive and required in-house expertise or qualified system integrators. Thankfully, advancements in technology (both hardware and software) have resulted in more capable, powerful, and flexible components. Presently, there is a wide choice of off-the-shelf products, like camera, lighting, and software, that system integrators can use to design systems with ease, while obtaining the same functionality at low cost. This has made the task of developing a vision system a relatively easier, simpler, and, most importantly, cost-effective.

Lastly, all manufacturing processes have some degree of variability. Machine vision technology is robust enough to compensate automatically for minor differences over time. However, the vision system may not be able to cater to major changes. For example, variations could be caused by ongoing modification to the production process or by substitutions of components and materials. In such cases, the application would need to be modified or even redesigned to cater to the ongoing modifications and new requirements.

4.4 Machine Vision Systems in Industry

Let us now look at different applications of machine vision for quality control in various industries.

The **automobile industry** was one of the earliest industries to adopt machine vision for quality control. Automotive assembly lines typically have hundreds of inspection points. These inspection points include in-process monitoring, individual part/component monitoring, and inspection at many critical points in the assembly line. Manufactured parts/components need to be monitored to ensure that their dimensions are as per specifications and aligned correctly for assembly purposes.

There are many challenges involved in designing systems for the automobile industry, such as inspection of multiple features of a part, multiple complex parts as well as inspection of different kinds of materials. Assembling large objects such as automotive bodies and their subassemblies require inspection of multiple features, such as surfaces, shapes, holes, slots, nuts, and studs.

Each part may have its own requirements solutions for quality control. Consider the part shown in Figure 4.1. As you can see there are many features to inspect. For example, the number and size of the holes must match specifications. The overall shape and dimensions must also match the specifications. Only then will this part fit into the subassembly/assembly correctly. To inspect feature by feature, multiple inspection cycles would be required, wherein one feature is inspected in one cycle. This process would be slow. It may be possible to design an efficient but more expensive solution that uses multiple cameras to make the several measurements simultaneously.

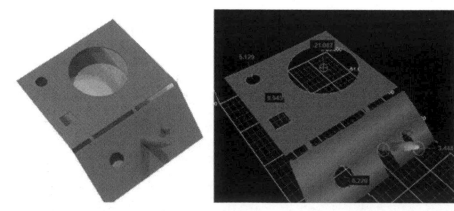

FIGURE 4.1
Sample automobile part.

Parts can vary in terms of materials and color and each of these may require separate inspections solutions to be developed. For instance, the illumination used would be different for different materials as the reflective power of the materials varies. Likewise, the image processing solution may have variations to handle the different requirements. All these would add to the inspection complexity and processing time.

Present-day automotive manufacturing processes are highly complex. More than any other industry, a vast majority of the parts/components are manufactured by third-party suppliers. To avoid problems during assembly and to achieve the high quality that is demanded by the customers, manufacturers and their suppliers must inspect and prevent defects during the entire production process.

Part traceability is another important application where parts are identified by decoding OCR and barcodes. Tracking of parts, assemblies, and final product helps to quickly identify the faulty defective parts in the production process. Preemptive corrective actions can be promptly taken to reduce defects and improve quality. Tracking helps to validate the regulatory compliance of the products and protects the company from statutory liabilities and financial risks. For example, there could be liabilities due to supply of faulty products, and recall and replacement of such defective products would be expensive.

Quality inspection can be used to find missing parts and for completeness checking. Color sorting is another application. For example, manufactured car handles need to be sorted on color basis to be fitted in the appropriate colored car. Machine vision can be used to inspect the size, shape, treads, etc., of car tyre. Different automobiles can have different parts depending on price range and machine vision systems can be designed to classify those automotive parts. Automobile manufacture is highly mechanized, and robots are extensively deployed to perform a variety of tasks, including assembly, painting, welding, parts transfer within the factory, as well as for tasks that involve segregation and sorting. Machine vision systems can be designed for robotic guidance.

It can be seen, that machine vision is used in the automobile industry for a wide variety of quality inspection tasks. Given the extremely competitive market for automobiles, flexible design with zero defects is essential to achieve high quality at low cost.

The **electronics industry** uses machine vision to inspect PCBs (printed circuit boards) as well as manufactured or bought out components. Integrated circuits (IC) have high levels of component density and integration. Machine vision technology has enabled visual inspection of even the smallest parts at real-time processing speeds. Vision systems provide high resolution along with high accuracy and reliability. Suppliers of electronic materials, active components, IC packages, passive components, and finished electronic equipment all use machine vision to drive high-quality production at lower costs.

The **pharmaceutical industry** has adopted machine vision in a major way to help guarantee product integrity and safety. Stringent regulatory standards exist for the pharma industry; hence quality automatically is a top concern because the liabilities for errors are very high. Machine vision not only helps to deliver high-quality products but also to achieve high productivity gains.

Drugs have a limited shelf life and drugs that have expired cannot be sold. Sometimes, an entire batch of drugs has to be recalled due to defective manufacturing. Tracking date and lot codes is therefore a very important application. Machine vision systems can be designed to satisfy the requirements of the extremely important application of accurate and reliable tracking and for verification of date and lot codes.

The pharmaceutical industry is one of the principal industries that uses color-based machine vision. Pills come in different colors. Color is often used to visually identify pills. For example, color checking is used for inspection of birth control tablets to ensure that the pills of the appropriate colors are present at specific locations of the blister package. Cap/seal, empty glass, liquid fill, blister seal, and box label inspections are other applications of machine vision.

Agriculture is experiencing an automation revolution. Agriculture is facing increasing global competition on the one hand while labor and operating costs are become higher. Farmers are turning to automation to help achieve improved yields at low operating costs. There are many types of automation technology in agriculture—different types of mechanized implements and tools—that help to make the physical tasks easier. Machine vision is one of the core technologies that helps to support precision and sustainable agriculture. For example, field robots are being extensively experimented and deployed for automatic harvesting, planting, growth monitoring, and weeding.

Machine vision technology has been found to be a scientific and effective tool for quality inspection, measurement, and evaluation of food grains, fruits, and vegetables. Sorting and grading of food products was principally done by the farmers and distributors using manual methods, such as visual quality inspections by human operators. Such processes are tedious, time-consuming, and error prone. Machine vision applications can be used to examine aspects such as appearance, color, texture, size, and visual appearance to assess food quality.

Presence of insect infestation and other blemishes can be detected using machine vision solutions. For example, dark spots in fruits and vegetables may indicate decay or the presence of worms. Color is an important aspect in detecting the maturity of fruits and vegetables; for instance, raw tomatoes are green, while ripe tomatoes are red. Some fruits or vegetables may be required to be a uniform single color. For other fruits, the presence of a second color may indicate maturity. For example, mangoes come in various colors—yellow, yellow-orange, and green, with some showing shades of red to pink, so for some varieties, the primary color of a ripe mango may be yellow but colors like orange and red can also be present.

Machine vision solution is applied in the **poultry industry** to inspect and grade eggs, as well to detect blemishes and cracks in the eggs. **Bakeries** use machine vision to determine the quality of baked goods and products. For example, the quality of bread can be determined based on the height and slope of the bread, while the quality of cookies can be determined by looking at their size, shape, and color.

Aquatic food processing is another area where machine vision technology is being applied. Machine vision systems help in the evaluation of size and volume, measurement of shape parameters, and quantification of the meat content present in the aquatic foods. Machine vision solutions are being developed for many functions such as sorting by species, sorting by size, sorting by visual quality attributes, automated portioning, and identifying defects in the species.

There are many more applications of machine vision solutions in various industries. The **textile industry** uses machine vision solutions to inspect the quality of the yarn and cloth produced. Machine vision systems are deployed to inspect the texture of the material as well as detecting the yarn evenness. The **tile and ceramic industry** uses machine vision for quality inspection of tiles for surface blemishes, cracks, and other irregularities, and to sort based on color and pattern. The **printing industry** uses machine vision for process control and print control. For example, the newspaper you read in the morning may be scanned by a machine vision system, and 100% inspection of the printed pages is used to check for print quality. Machine vision solutions help to detect defects in the print pages, such as color variations, color deviation, smearing or streaks, misprint, double print, spots, etc.

Machine vision solutions in the industry are very versatile. Some are established solutions, while others are still in the R&D stage. There is no dearth of applications for which machine vision can be effectively applied to improve quality control and optimize the production process.

4.5 Categorization of Machine Vision Solutions

We have categorized the quality control applications in the industry under the following broad headings:

- Dimensional measurements
- Presence/absence
- Character inspection (OCR, barcode)
- Profile inspection
- Surface inspection
- Robotic guidance

This classification is based on the type of quality inspection that is carried out. A large part of the control activities would fall under one or more of these categorizes or combinations of these categorizes. There will always be exceptions, and such applications may have unique requirements that may not fit in any of these categorizes.

We will now discuss the type of quality inspection that is carried out under each category and use examples to illustrate the applications.

4.5.1 Dimensional Measurement

Dimensional measurements are carried out to determine whether parts/ products have been manufactured as per specifications. This includes checking for correctness of length, width, height, thickness, diameter of a circular part or hole, etc. As mentioned earlier, a part must be produced according to specified dimensions for it to fit correctly in the subassembly or assembly. A certain level of tolerance is allowed with respect to these specifications. For example, the desired length of an object is 3 centimeters, but a tolerance of −5% to +5% may be allowed. The following are examples of dimensional measurement:

- Measuring the length and breadth of parts/objects
- Measuring the diameter of circular objects
- Measuring the angle at the edges of objects
- Measuring the position of labels or characters
- Measuring the width of parts/sheets/films

In dimensional measurement, it is often necessary to measure multiple dimensions like length and breadth for the same part. If the part meets the criteria specified, then it is deemed to have passed the quality check and is accepted. Depending on the defect, some of the parts which fail to meet the quality requirement may be rejected outright. Others may undergo modification or rework and again go through the process of quality inspection.

In the manual approach, the dimensions of parts and products are measured with micro gauges or calipers or checked with inspection jigs to ensure that there is no variation in accuracy. Other tools like profile measurement systems and optical comparators are used to carry out dimension measurement. The above methods are simple to operate but are time-consuming, expensive, and prone to errors. It is also impossible to do 100% measurement checking. Only samples from a production batch are checked.

With machine vision, multiple dimensions can be obtained from captured images. It is easy to measure the dimensions of individual cross sections of parts and products based on captured images. It is also possible to measure angles or circle roundness along with the lengths of various sections.

The quality control decisions, such as pass/fail, and the numerical data can be saved and used for traceability management or process improvement. Let us look at some examples from industry.

4.5.1.1 Dimensional Measurement of Oil Seal

Oil seals are used in automobiles to prevent leakages and to provide protection from dust and other grime. It, therefore, is essential that the dimensions of the part are accurate. Figure 4.2 shows a sample oil seal.

It can be seen from the figure that the oil seal consists of two rings. A machine vision system can be developed to measure and verify the inner and outer diameters. The measured values must match with the numbers given in the specifications within allowable tolerance limits. Oil seals that satisfy the specifications would be accepted; those that don't would be rejected or sent for rework.

4.5.1.2 Dimensional Measurement of Reed Valve

Reed valves are a type of check valve used to restrict the flow of fluids. They open and close based on the pressure of the liquid but allow flow in one direction only. Reed valve are used in automobile air-conditioning systems, as well as in the EGR (exhaust gas recirculation) systems of a wide variety of diesel engines. For these valves to work correctly, it is essential that their dimensions match requirements. Figure 4.3 shows a sample reed valve with dimensions measured using a machine vision system. The figure shows the reed value with the measurements obtained for the various holes/cavities. For circular holes, diameter is measured as shown in the figure. It can be noted that there are minor variations in the values obtained; the diameters

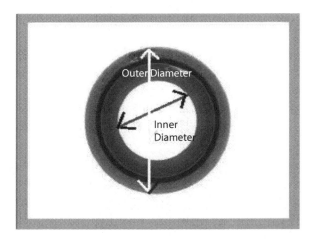

FIGURE 4.2
Dimensional measurement—oil seal.

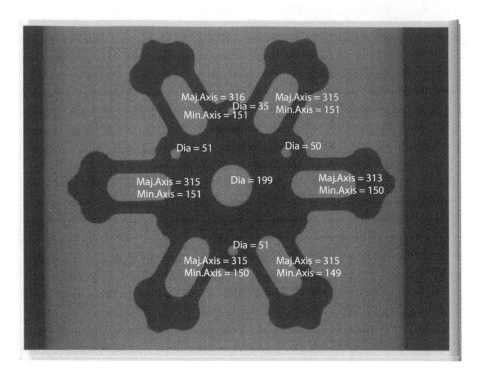

FIGURE 4.3
Dimensional measurement—reed valve.

of the three small circles are 51, 50, and 51. Likewise the maximum axis is 315 or 316, except in one case where it is 313. The dimensional measurements of the reed valves can be matched with the design specifications to determine if the dimensions are accurate and within the permitted tolerance values.

4.5.2 Presence/Absence Inspection

Another key area where machine vision systems are used is for presence/absence checking in parts and packaging. The parts or components can be inspected and counted as they move on the production line. Inspection can also be carried out to check the quantity of parts or objects at a workplace. There are many types of presence/absence inspections, such as the following:

- Counting bottles packaged in boxes or containers
- Checking packaging to verify presence of accessories that should be included
- Checking for the presence of labels and codes
- Checking ICs for presence/absence of electronic components

- Checking for the presence of screws and washers used for securing different parts
- Checking for the presence/absence of tablets
- Checking for the completeness of tablets packaging

Manual verification by human operators requires a high degree of attention and concentration and such inspection is always prone to errors due to human fatigue or oversight. The physical conditions present for doing inspection as well as the experience of the inspectors are important criteria for the success of inspections. Variations in the accuracy of human visual inspection due to skill differences among workers are also likely.

Machine vision solutions can be used to implement 100% quality check. Many methods can be used to implement presence/absence checks, such as binary processing and blob analysis. Binary processing converts data into two levels—black and white. Pattern matching and template matching can be used to detect presence and for counting. In matching, a standard template that shows the desired image is stored. This is compared with the target image to detect presence/absence and do the counting. Let us now look at some sample industrial examples.

4.5.2.1 Blister Pack Inspection

Blister packs are used to package pills or tablets in the pharmaceutical industry. The packaging protects tablets from dust, dirt, humidity, and other external factors. A blister pack should have a specified number of pills, for example, 10 or 12 tablets. The count can vary depending on the tablet size as well as standard dosages prescribed by doctors.

A machine vision system can be developed to inspect the blister packs for the presence/absence of tablets in the pack as well for completeness—whether all tablets are present. Figure 4.4 shows samples images of blister packs.

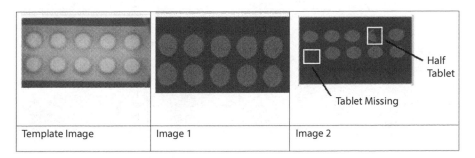

| Template Image | Image 1 | Image 2 |

FIGURE 4.4
Presence/absence inspection of a blister pack.

Figure 4.4 shows the standard template and the captured images of sample blister packs. Image 1 shows a blister pack where all the 10 tablets are present as required. Image 2 shows a blister pack where two tablets are not okay; one tablet is missing while another is only a half tablet.

4.5.2.2 Bottle Cap Inspection

In this example, bottles caps are inspected for completeness and absence of any part. Figure 4.5 shows two images of bottle caps that are to be inspected.

It can be observed that the bottle cap consists of a top cover or cap, a plastic ring below, and a sealing ring that binds the cap and the ring. It can be seen, that in the first picture, bottle cap is correct, with all parts present. In the second picture, the sealing ring is missing.

4.5.3 Character Inspection

Character inspection is used to verify the characters that are printed on labels, parts, or products, as in the following examples:

- Verifying if barcodes are printed correctly
- Verifying QR (quick response) codes for correctness
- Reading and verifying part numbers or model numbers
- Checking expiration dates on drug packaging
- Checking expiration dates on food containers

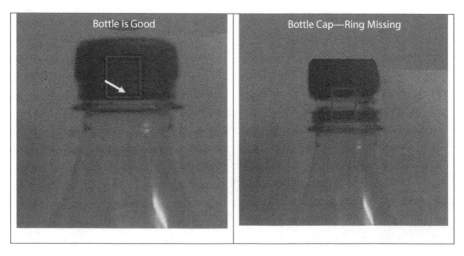

FIGURE 4.5
Presence/absence inspection of a bottle cap.

Barcodes and QR codes are used to visually represent data about a part. The information encoded can be either numeric or text or both. The visually represented data is machine-readable. Hence, machine vision systems can be developed to read the codes and verify their correctness. Such codes can also be used for part verification or tracking.

The codes can be printed or stamped on cartons, metal parts, products, etc. Barcodes represent data by using parallel lines of varying widths and by different spacings between the parallel lines. These are also referred to as one-dimensional (1D) code. A QR code is another popular code that contains information that is machine readable. It is a two-dimensional code, consisting of black-and-white patterns. It is capable of storing alphanumeric content and the information is available in both horizontal and vertical directions. Figure 4.6 shows a barcode and a QR code.

Character inspection is another important application of machine vision technology. Labels indicating product numbers, expiration date, etc., printed on parts and products can be read and interpreted. Machine vision solutions automate reading of these labels, and the information they provide can be analyzed, interpreted, and stored.

Machine vision solutions for code inspection are used to verify presence of the code as well as to check for accurate positioning, formation, and readability. Such systems can automatically identify and reject containers or packages with missing, incorrect, or unreadable codes. Noncompliance can be traced and acted on.

4.5.3.1 Label and Barcode Inspection

Figure 4.7 shows the image of an assorted packet of biscuits.

A machine vision system can be designed to read and verify the position of the text as well as its contents. The text as shown—"BRAND X" and "200 g"—and the barcode on the right-side corner of the packet can be read and verified for correctness.

(a)

(b)

FIGURE 4.6
Sample images of (a) barcode and (b) QR code.

FIGURE 4.7
Sample image for label and barcode inspection.

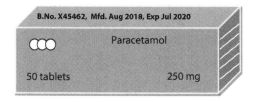

FIGURE 4.8
Sample image for drug pack inspection.

4.5.3.2 Drug Pack Inspection

Figure 4.8 shows the picture of a sample blister pack cover. One or more sheets of tablets are packed within the cover. A blister pack cover is marked with product details as well as manufacturing and expiration dates. This information is normally printed on the cover as well on each of the tablet sheets.

From the figure it can noted that the following information is printed on the blister pack cover:

- Name of the tablet (Paracetamol)
- The total number of tablets in the pack (50)
- The quantity of the chief ingredient in the drug (250 mg)
- Batch number, manufacturing, and expiry dates

A machine vision system can be developed to verify all the above information. Figure 4.9 shows a drug pack which is being verified for correctness. It is evident that the following information printed on the cover is incorrect:

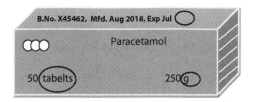

FIGURE 4.9
Sample image of drug pack with missing information.

- The word "tablets" is misspelled as "tabelts."
- The quantity of the chief ingredient in the drug has been wrongly printed as "250 g" instead of "250 mg."
- The year is missing from the expiry date.

4.5.4 Profile Inspection

Profile inspection checks for the presence of flaws and defects on the surface of parts or products. Examples include:

- Checking for stains on cloth
- Checking for burrs on machined metal parts
- Checking to see if edges are smooth or rounded
- Checking for defective LEDs

Examples of profile inspection are explained below.

4.5.4.1 Profile Inspection of Spline Gear

Spline gear is an automobile component that is used in cars for gear adjustment. Bad quality can lead to failures and accidents. Figure 4.10 shows images acquired using a camera with frontlighting and backlighting.

While the lighting depends on the product and the type of inspection required, in general, backlighting is more suitable for profile inspection while frontlighting is appropriate for verification of presence/absence in objects. As seen from the figure, backlighting emphasizes the edge profile of the spline gear. A sample machine vision setup and the output user interface are shown below in Figure 4.11.

In the quality inspection process, the image of the spline gear to be verified is captured. The captured image is matched with a standard template (correct image) of the spline gear. For spline gears that pass the inspection test, the OK button lights up. For defective parts, the NOT OK button turns on. It can be noted, that the sample setup shown in the figure, has provision for quality inspection of different types of gears as indicated by the buttons at the left-hand side of the figure.

Image taken using frontlighting   Image taken using backlighting

FIGURE 4.10
Spline gear inspection—images of sample spline gear.

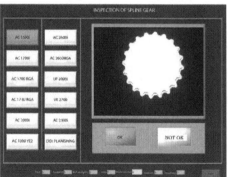

FIGURE 4.11
Spline gear inspection—machine vision setup.

4.5.4.2 Profile Inspection for Packaging Integrity

Bottles are packed in crates or boxes. Such crates can be checked for completeness—whether all bottles are present. The individual bottles can be checked to see if liquid is present, and whether they are filled to the correct level. Damages to the crate as well as defects to the bottles can be checked. Figure 4.12 shows the image of a bottle which is being checked for fill level. Two buttons OK and NOT OK are shown. If the fill level in the bottle is correct, the OK button lights up, else the NOT OK button is turned on.

Full crates can be checked for completeness and defects and rejected if there are defects such as missing bottles, bottles with incorrect fill levels, and missing or defective bottle caps or covers. Machine vision systems can detect the profiles with high micron accuracy. However, profile inspection of irregular shapes requires specifically designed solutions. The speed of inspection must also match the speed of packaging to ensure that all crates are inspected correctly.

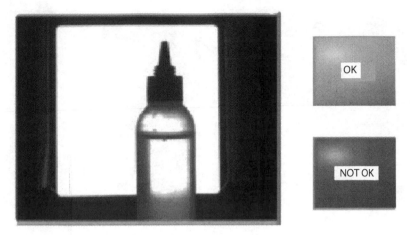

FIGURE 4.12
Profile inspection—checking bottle fill levels.

4.5.5 Surface Inspection

Visual inspection of finished products is one of the basic applications for quality control in many industries. The ability of the products to perform their expected functions is important. However, customers also look for visual appeal. For example, customers expect a mobile phone to be visually appealing in terms of color, shape, and other aspects. Traditionally, inspection was primarily a manual process done by human operators. Machine vision technology can replace manual inspection with an efficient, accurate, and reliable process that can provide 100% inspection of parts/products. The solution has to be designed depending on the product and the type of inspection(s) required. The parameters of the inspection vary depending on the type of object to be inspected.

One example, which we discuss here, is from the food industry where quality is assessed on basis of appearance, color, texture, size, and other visual parameters. Figure 4.13 shows the sample and processed image of an apple. The defective part is highlighted in white. This application can help to sort and segregate good and bad apples.

The parameters for comparison can vary for different fruits as the color and shape of each type of fruit is different, as also its desired appearance. For example, yellow color may indicate a ripe banana or mango but would not apply to apples where the required color is red.

Another example comes from the automobile industry. Car tyre, which are fitted on the car wheels on rims, are examined for cracks which may be present on the surface. Figure 4.14 shows images of a good tyre (at left) and one with surface cracks (at right). The crack is highlighted to reveal its presence.

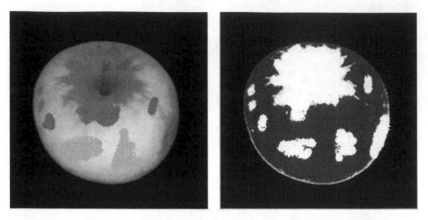

FIGURE 4.13
Surface inspection—image of an apple and the processed output.

FIGURE 4.14
Surface inspection—image of a tyre and its processed output.

This is an important safety consideration, as tyre cracks can lead to faulty grip and road accidents.

4.5.6 Robot Guidance

Application of machine vision for robotic guidance is a rapidly developing area. Robots can be used for a variety of tasks, such as location of parts, orientation of parts, arranging parts on/off production line, packaging parts, placing parts on a work shelf, or removing parts from bins. Cameras, vision algorithms, and sensors are used to guide robots. A single robot can be used for multiple tasks or to take care of multiple machines. Algorithms are constantly being improved. For example, vision can help robots avoid collision with other robots. They can stop if there are other objects in their motion path.

4.6 Summary

We started with an overview of quality control and the need to establish a quality management system. Whereas quality assurance focuses on defect prevention, quality control aims to identify defects, that is, products that fail to meet the product specifications. QA/QC is the combination that ensures products and services are designed and manufactured to meet the customer expectations. We saw how machine vision is aptly suited for quality control and explored the advantages of accuracy and speed it provides at optimal cost.

We then looked at various industrial applications that use machine vision for quality control. We broadly classified these applications based on type of quality check that is carried out, namely, dimensional measurement, presence/absence, character inspection, profile inspection, surface inspection, and robotic guidance. We looked at various examples from several industries for each of these classifications. Many industrial applications fall within these categories but there are always exceptions.

Exercises

1. What are the key benefits of using machine vision for quality control? Discuss in detail using examples.

2. Compare and contrast manual inspection with automated inspection.

3. Discuss two examples where machine vision cannot be used.

4. Discuss an example of a machine vision application where, in your opinion, color imaging is mandatory.

5. Consider an application where 3D imaging would be beneficial. Explain the application giving the limitations of using 2D and the advantages of using 3D imaging.

6. What is the difference between product inspection and packaging inspection? Provide examples in your discussion.

7. What type of lighting would you use for reflective parts? Explain using examples.

8. When would you decide to modify an existing machine vision system? Use examples to explain why and how you would redesign an existing machine vision system. Also explain when and why you would develop a totally new system.

9. "Agriculture is experiencing an autonomous revolution." Justify the statement expanding on examples given in this chapter.

10. Consider an application that performs quality inspection of an automobile door. Prepare a complete checklist of the inspection items and explain the type of inspection you would perform. Differentiate between inspections for which you would use machine vision and those for which you would not, and explain your answer.

11. Consider an application that performs quality inspection of a loaf of bread made in a bakery or factory. Prepare a complete checklist of the inspection items and explain the type of inspection you would perform. Differentiate between inspections for which you would use machine vision and those for which you would not, and explain your answer.

12. Consider an application that performs quality inspection on a packaged bundle of white A4 paper. Prepare a complete checklist of the inspection items and explain the type of inspection you would perform. Differentiate between inspections for which you would use machine vision and those for which you would not, and explain your answer.

13. Consider an application that performs quality inspection of a mug of coffee served in a restaurant. Prepare a complete checklist of the inspection items and explain the type of inspection you would perform. Differentiate between inspections for which you would use machine vision and those for which you would not, and explain your answer.

14. Continuous rolls like paper, plastics, or other materials use high-speed machine vision web inspection system. Explain web inspection using a sample application.

For questions 15–20, discuss examples not given in this chapter.

15. Describe an industrial application where machine vision can be used for dimensional measurement.

16. Describe an industrial application where machine vision can be used for checking presence/absence.

17. Describe an industrial application where machine vision can be used for profile inspection.

18. Describe an industrial application where machine vision can be used for surface inspection.

19. Describe an industrial application where machine vision can be used for robotic guidance.

20. Describe an industrial application where machine vision can be used for label inspection.

5

Digital Image Processing for Machine Vision Applications

Development of machine vision applications is a complex task that involves—understanding requirements, selection of hardware, selection of software, and last but not the least the selection of the appropriate processing methodology to obtain the most accurate and optimal solution.

In Chapter 2, we covered the basics of digital images. We learned that digital images can be of many types: color, grayscale, or black-and-white; as well as 2D, 3D, 4D, and other multidimensional images. Brightness, contrast, resolution, and color are some of the key characteristics that define a digital image. We also know that digital images come from various sources including medical imaging (X-rays, ultrasound, etc.), satelliteimaging, graphic software (CAD, SolidWorks, etc.), and photography. Digital images can also be obtained by scanning drawings, printed materials, books, etc. Each type of image has its own unique characteristics and processing requirements. In Chapter 3, we saw how machine vision cameras are used to capture an image. We also learned the importance of matching cameras with appropriate lenses and choosing the correct illumination to highlight the required features in images.

We know that image processing is the mechanism for extracting information from a digital image. Many methods and techniques are available for processing these digital images. Many algorithms have also been developed, with variations, that use these techniques for image processing. Our aim in this chapter is to discuss some of the important methods and techniques that are typically used in building machine vision systems for quality control. This does not, however, presuppose that other digital image processing techniques do not apply. As repeatedly mentioned, the choice of technique depends on the data being analyzed and the outcome desired by the application. We encourage the readers to look up the several books titles and other resources that have been provided on image processing in the reference section. Referring to them would help to strengthen the understanding of the concepts and provide useful access to other techniques.

Many of the quality control applications in industry use 2D images for inspection. Hence, this chapter covers image processing of 2D images; 3D image processing is an emerging trend and is discussed in Chapter 7. In Chapter 2, we covered certain basic operations that can be used to transform an image, namely, point operations, thresholding, geometric and affine transformations, and image interpolation. We will be referring to these processes in this chapter. We shall discuss 2D digital image processing under the broad headings outlined in Chapter 2—preprocessing, segmentation, and object detection/recognition.

5.1 Preprocessing

Preprocessing is an important step in image processing. Raw image data from cameras may have a variety of problems such as noisy or poor edge points, blurred focus, incorrect illumination, etc. For example, incorrect lighting can introduce shadows that obscure the texture or features of the image.

Preprocessing is carried out to improve the quality of the image and emphasize the features required by the application. For profile inspection, preprocessing can be used to sharpen features such as edges and boundaries of the objects in the image. While the inherent content of the image is not altered, specific features can be accentuated so that they can be easily detected. This improves the quality of feature extraction which is directly related to the quality of results obtained from image analysis. Image preprocessing is analogous to the mathematical normalization of a data set, which is a common step in many feature descriptor methods. Some of the key preprocessing operations are as follows:

- Filtering
- Scaling/subsampling
- Histogram generation

Table 5.1 lists some of the common image preprocessing operations, with examples from each of the three operations given above, illustrating both differences and commonalities among these image preprocessing steps. Our intent here is to illustrate rather than prescribe or limit the methods chosen.

5.1.1 Image Filtering

Filtering is a technique that is used to modify or enhance an image. Image processing operations implemented with filtering include smoothing, sharpening, and edge enhancement. Filtering is a neighbored operation, in which the value of a given pixel in the output image is determined from the values of the pixels in the neighborhood of the corresponding input pixel in the input image. You may recall that we learnt about 4-connected neighborhood and the 8-connected neighborhood in Chapter 2, where the value of the

TABLE 5.1

Common Image Preprocessing Operations

Preprocessing for	Filtering	Subsampling/ Scaling	Histogram Generation
Illumination rectifications	√	√	√
Blur and focus rectifications	√	√	√
Noise removal	√	√	√
Thresholding			√
Edge enhancements	√	√	√
Segmentation		√	√
Region processing and filters	√	√	√
Color space conversions		√	√

output image pixel is obtained by applying some processing methodology to the values of the pixels in the neighborhood of the corresponding input pixel. This process helps to enhance certain features while preserving the other features.

A digital image often contains noise or distortions, which introduces erroneous pixel values. Images generally contain objects that have homogenous intensity, that is, roughly the same intensity as their neighbors. This is used to filter out noise. Let us consider an image acquired using a camera, which is noisy, that is, the intensity levels are altered a little bit. Consider a 3×3 matrix within the image with the following values:

$$V = \begin{pmatrix} 5 & 3 & 4 \\ 4 & 3 & 4 \\ 4 & 5 & 4 \end{pmatrix}$$

The values in the matrix should have all been 4, but due to noise the intensity is varied. One method of filtering noise is to adjust by taking the local average of the neighborhood. The value at the center of the matrix is 3. In the output matrix, the average value considering the 4-neighborhood would be

$$3 + 4 + 4 + 5/4 = 4$$

The value 4 is substituted in the output image, reducing noise in the image. However, if one of the values were to be substantially higher, say 15, then taking the average would not help.

Linear filtering is normally carried out by an operation called "convolution." In this method, the output pixel is the weighted sum of the input neighboring pixels. The matrix of weights is called the "convolution kernel," which is the filter. Let the convolution matrix, kernel K be defined as a 3×3 matrix as follows:

$$K = \begin{pmatrix} a & b & c \\ d & e & f \\ g & h & i \end{pmatrix}$$

The first step of convolution is to flip the columns and rows of the kernel matrix. The flipped matrix is obtained by rotating the values about the center element of the matrix. Let us represent the flipped matrix of K by K_f:

$$K_f = \begin{pmatrix} i & h & g \\ f & e & d \\ c & b & a \end{pmatrix}$$

If the kernel chosen is a symmetric matrix, then flipping rows and columns will return the original kernel matrix.

The next step is to slide this flipped matrix through the input image by placing the center of K_f over each image pixel. When we do this, we replace the pixel value by multiplying kernel elements with the pixel values directly below them and then taking the sum. Let us assume that at a given instance I_m is the image matrix. The kernel K_f is placed over the image coordinates (x, y):

$$I_m = \begin{pmatrix} p1 & p2 & p3 \\ p4 & p5 & p6 \\ p7 & p8 & p9 \end{pmatrix}$$

Then in the transformed image, at the pixel position (x, y) the pixel value is

$$i \times p_1 + h \times p_2 + g \times p_3 + f \times p_4 + e \times p_5 + d \times p_6 + c \times p_7 + b \times p_8 + a \times p_9$$

From the above equation, we can derive that each image pixel value is replaced by the weighted sum of neighboring pixels with weights defined by the flipped kernel elements over them. If I denotes the input image matrix, then the transformed or output image obtained by convolving with K is denoted as $(I \otimes K)$.

Convolution is represented by an equation:

$$(I \otimes K)[x,y] = \sum u \sum v\, I[x-u, y-v]\, K[u,v]$$

Convolution is commutative, that is,

$$A \otimes B = B \otimes A$$

Convolution is associative as represented by

$$(A \otimes B) \otimes C = A \otimes (B \otimes C)$$

Convolution also satisfies the distributive law:

$$A \otimes (B + C) = A \otimes B + A \otimes C.$$

We will now discuss some of the filters that are commonly used for preprocessing images.

5.1.1.1 Normalized Box Filter

If all the elements of a kernel are given unit values, convolving it with image would mean replacing the pixel values with the sum of its neighbor in $K_{height} \times K_{width}$ window, where kernel size is given by height and width. If each element in the kernel is now divided by the kernel size, then the sum of all elements will be 1—the normalized form. Such a convolution would result in pixel values being replaced by mean or arithmetic average of neighboring pixels in the $K_{height} \times K_{width}$ window. Taking the average would mean reduction in sudden changes in intensity values between neighboring pixels. This filter helps to achieve smoothness in images.

Normalized box filter is otherwise known as moving average filter. Although the moving average filter is simple and fast, it has two drawbacks. First, it is not isotropic (i.e., circularly symmetric) as it smooths further along diagonals than along rows and columns. Second, weights have an abrupt cutoff rather than decaying gradually to zero, which leaves discontinuities in the smoothed image.

5.1.1.2 Gaussian Filter

Gaussian filters use weights specified by the probability density function of a bivariate Gaussian, or normal, distribution with variance σ^2, that is,

$$w_{ij} = \frac{1}{2\pi\sigma^2} * e^{-(x^2+y^2)/2\sigma^2}$$

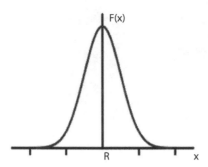

FIGURE 5.1
Gaussian distribution.

for $x, y = -[3\sigma],...,[3\sigma]$, for some specified positive value for σ^2. The weight is represented by w_{ij}; (x, y) is the pixel coordinates, and $[3\sigma]$ represents the "integer part" of 3σ. Limits of $\pm3\sigma$ are chosen because Gaussian weights are negligibly small beyond them. Note, that the divisor of $2\pi\sigma^2$ ensures that the weights sum to 1 (approximately), which is a common convention with smoothing filters. Gaussian distribution is shown in Figure 5.1.

In Figure 5.1, we can see that the values near the reference points R, are more significant. This same Gaussian distribution is achieved in a 2D kernel with the reference point being the matrix center. As opposed to the normalized box filter that gives equal weight to all neighboring pixels, when calculating sums a Gaussian kernel gives more weight to pixels near the current pixel and much lesser weight to distant pixels. It may be noted that the Gaussian filter can also be used to simultaneously smooth and interpolate between pixels, provided that $\sigma^2 \geq 1$. Hence, it can be concluded that this kind of filter, which operates in a larger neighborhood, will be more effective at reducing noise but will also blur edges. Gaussian filters overcome the drawback of moving average filters where weights decay to zero.

5.1.1.3 Bilateral Filter

While Gaussian filter gives more accurate results when compared to box filter, both the filters smoothen the edge pixels, thereby diminishing the intensity value. This can be overcome by using a bilateral filter. Bilateral filtering smooths images by means of a nonlinear combination of nearby image values. It combines intensity values on basis of spatial location or domain (geometric closeness or neighborhood), as well as on range (pixel intensity values). Near values are preferred to distant values in both domain and range.

The bilateral filter is also defined as a weighted average of nearby pixels, in a manner very similar to Gaussian convolution. The difference is that the bilateral filter takes into account the difference in value with the neighbors to preserve edges while smoothing. The key idea of the bilateral filter is that for a pixel to influence another pixel, it should not only occupy a nearby location but also have a similar value. In order words, the rationale of bilateral filtering is that two pixels are close to each other not only if they occupy nearby spatial locations but also if they have some similarity in the photometric range, that is, pixel intensity. The bilateral filter, denoted by $BF[I]_p$, is defined by the following equation:

$$BF[I]_p = \frac{1}{W_p} \sum_{q \in s} G_{\sigma s}\left(\|p-q\|\right) G_{\sigma r}\left(I_p - I_q\right) I_q$$

where W_p is a normalization factor and is defined by

$$W_p = \sum_{q \in s} G_{\sigma s}\left(\|p-q\|\right) G_{\sigma r}\left(I_p - I_q\right)$$

Parameters σs and σr will measure the amount of filtering for the image I.

$BF[I]_p$ is a normalized weighted average where $G_{\sigma s}$ is a spatial Gaussian that decreases the influence of distant pixels; $G_{\sigma r}$ is the range Gaussian that decreases the influence of pixels q with an intensity value different from I_p. Some of the key merits of bilateral filter are as follows:

- Its formulation is simple: each pixel is replaced by a weighted average of its neighbors. This aspect is important because it makes it easy to acquire intuition about its behavior, to adapt it to application-specific requirements, and to implement it.
- It depends only on two parameters that indicate the size and contrast of the features to preserve.
- It can be used in a non-iterative manner. This makes the parameters easy to set since their effect is not cumulative over several iterations.

5.1.1.4 Comparison of Filter Techniques

A sample of input image and the respective filter outputs are shown in Figure 5.2. Note that the Gaussian filter blurs the image, the box filter smooths the image, and the bilateral filter sharpens the edge points. Hence, the type of filtering technique needed should be chosen depending on the requirement.

Bilateral filters provide noise reduction while maximally preserving edges and peaks. Such filters are suitable for profile inspection. Gaussian filters are computationally faster, providing good noise reduction but smooths edges. Gaussian filters can therefore be used for surface inspection. If the requirement is only to reduce noise, then the optimal solution is the normalized box filter. For optimum computational cost, kernel size should ideally be less than or equal to 7.

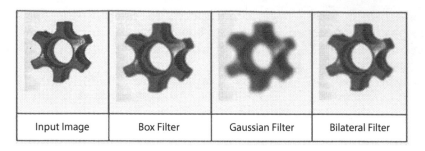

| Input Image | Box Filter | Gaussian Filter | Bilateral Filter |

FIGURE 5.2
Sample images obtained from different filtering techniques.

In summary:

- Filters are used to reduce noise and smooth or enhance edges of objects in images.
- Filters reevaluate the value of every pixel in an image. For a particular pixel, the new value is based on pixel values in a local neighborhood and a window centered on that pixel.
- Filters are considered linear if the output values are linear combinations of the pixels in the original image.
- Linear filters provide noise reduction but smooth edges.
- Nonlinear filters can provide noise reduction without blurring the edges and can detect edges at all orientations simultaneously.
- Linear filters are computationally faster in comparison to nonlinear filters

5.1.2 Subsampling/Scaling

Scaling was discussed in Chapter 2. Scaling is a type of geometric (affine) transformation that is used to resize images. Images are scaled to maintain a standard image size for further processing. Input images from different sources may be of different sizes and scaling is done to bring them to a standard size. Images can be scaled up or scaled down. Scaling can lead to loss of overall image quality. For instance, if the image is scaled down to a smaller size and then resized to the original size, the image will be a lot less clear.

Scaling down an image reduces the size of the image and hence reduces the image computation or processing time. For example, in profile inspection, images are scaled down while retaining the features necessary for inspection. Scaling up is done to enhance the finer details in the image. In surface inspection, for example, scaling up helps to find defects on the surface of the object.

Scaling can be carried out using interpolation. We discussed interpolation—nearest-neighbor, bilinear, and bicubic in Chapter 2. All of these methods can be used to for scaling of images. Nearest-neighbor is the simplest and perhaps fastest way of doing scaling. Nearest-neighbor and bilinear interpolation can introduce jagged edges when scaling. Bicubic interpolation produces smoother images but would take more time to execute. Each method comes with advantages and disadvantages and selection of the appropriate one depends on application requirements.

Nearest-neighbor interpolation is the simplest approach to interpolation. Rather than calculating an average value by some weighting criteria or generating an intermediate value based on complicated rules, this method simply determines the "nearest" neighboring pixel and assumes its intensity value. In other words, nearest-neighbor is the most basic method and requires the least processing time of all the interpolation techniques because it considers only one pixel—the closest one to the interpolated point. This has the effect of simply making each pixel bigger. The nearest-neighbor method cannot be used for high-resolution zooming.

Bilinear interpolation considers the closest 2 × 2 neighborhood of known pixel values surrounding the unknown pixel. It then takes a weighted average of these four pixels to arrive at its final interpolated value. This results in much smoother looking images than nearest-neighbor interpolation.

Figure 5.3 compares the result obtained using nearest-neighbor and bilinear interpolations for better understanding.

Bicubic interpolation goes one step beyond the bilinear method by considering the closest 4 × 4 neighborhood of known pixels; for a total of 16 pixels. Since these are at various distances from the unknown pixel, closer pixels are given a higher weight in the calculation. Bicubic interpolation produces noticeably sharper images than the previous two methods and is perhaps the ideal combination of processing time and output

Nearest-Neighbor

Bilinear

FIGURE 5.3
Images obtained with bilinear and nearest-neighbor interpolation.

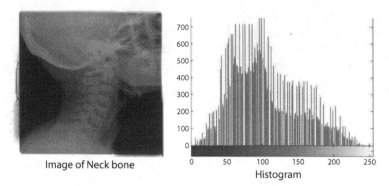

FIGURE 5.4
Image of a neck bone and histogram.

quality. For this reason, it is a standard in many image editing programs, printer drivers, and in-camera interpolation.

There are many other types of interpolation techniques that can be used for scaling. These include Fourier-based and edge-directed interpolation. Readers are encouraged to use the resources given in the References section for additional reading.

5.1.3 Histogram

Histograms of images are pictorial representations, showing the number of pixels at each different intensity value that is found in the images. In an image histogram, the *x*-axis shows the gray level intensities and the *y*-axis shows the frequency of these intensities.

Consider the image of the neck bone and its histogram shown in Figure 5.4. The *x*-axis of the histogram shows the range of pixel values present in the grayscale image: 0–255, which is 256 levels or shades of gray. The *y*-axis represents the count of pixels for each of the intensity values. Histograms are used to perform the frequency analysis of images and for adjusting the brightness/contrast of images.

5.2 Image Segmentation

We know that image analysis is carried out to obtain information that is used for automation in machine vision. Why is image analysis a difficult process? To get information from an image it is necessary to meaningfully understand and interpret the image.

FIGURE 5.5
Image of a bowl of fruit.

Consider the bowl of fruits shown in Figure 5.5. How does the human brain interpret this? To explain the process in a simplified way, we would look at the image and first try to differentiate between the background and the objects. Then we would try to understand what each of these objects is. We would probably think, "I can see a banana; I can recognize it from the shape, size, and color." Similarly, we would probably recognize the apple from the round shape, red color, and textural markings. In this manner, we would identify each of the individual fruits, then the plate, and finally conclude that what I have in front of me is a plate of fruits.

Image segmentation is that process that tries to partition an image into its several parts, in order that each of the parts can be analyzed and interpreted separately. Therefore, one of the aims of image segmentation is simplification, that is, partitioning an image to make it easier to analyze. However, it is important that the partitions are meaningful. Partitioning is carried out based on shape, size, intensity, texture, etc. For example, pixels that are near to each other and share the same pattern or intensity are grouped into a single object. Likewise, if the pattern or gradient changes between two pixels, then it probably implies that the two pixels are not of the same object. So far, most of our understanding of a digital image has been at the pixel level. It is only at this stage that we look at grouping pixels to make meaningful partitions or regions.

There are many methods for image segmentation. These include thresholding-based segmentation, edge-based segmentation, region-based segmentation, clustering methods, graph-based methods, pixel-based methods, and hybrid methods. Each of these methods has its own merits and disadvantages and the choice of methodology depends on the specific application requirements. Here we discuss in some detail the commonly used methods enumerated above. The Reference section lists resources for exploration of other methods.

5.2.1 Threshold-Based Segmentation

One of the simplest approaches to image segmentation—called thresholding— is based on the intensity levels of the pixels. Readers may remember that we have covered thresholding in Chapter 2.

To recall, threshold-based techniques classify images into two classes and work on the postulate that pixels belonging to certain range of intensity values represent one class and the rest of the pixels in the image represent the other class. Thresholding can be implemented either at a local or global level. Global thresholding distinguishes object and background pixels by comparing the pixel intensity with the chosen threshold value. The pixels that pass the threshold test are considered as object pixels and are assigned the binary value 1 and other pixels are assigned binary value 0 and treated as background pixels. Threshold-based segmentation techniques are inexpensive, computationally fast, and can be used in real-time applications with aid of specialized hardware. Thresholding is expressed using the following equation:

$$g(x, y) = \begin{cases} 1 \text{ for } i(x, y) >= t \\ 0 \text{ for } i(x, y) < t \end{cases}$$

where $g(x, y)$ is the output image; $i(x, y)$ is the input image and t is the threshold value.

Local thresholding is also known as "adaptive thresholding." In this technique the threshold value varies over the image depending on the local characteristic of the subdivided regions in the image. The image can be divided into sub-images and the thresholding value can be different for each subimage. Histograms, explained in the previous section, can be used to determine the threshold value. For instance, a histogram could have the pixel intensities clustered around two well-separated values. Thresholding value could be decided based on the peak values.

Merits:

- Computationally inexpensive
- Fast and simpler to implement
- Can work in real-time applications

Demerits:

- The spatial information of the image is neglected.
- The method is highly noise-sensitive.
- Selection of threshold value is crucial.
- Can result in over or under segmentation.
- May lead to pseudo edges or missing edges.

5.2.2 Edge-Based Segmentation

These types of segmentation methods are otherwise called discontinuity-based approaches and are based on the principle of intensity variations among the pixels. For example, boundaries exist if the image has two or more objects, and this can be used to segment the image. The boundaries of the objects form the edges. Boundaries are detected using discontinuities in pixel intensity. Significant and abrupt changes in the intensity levels among neighboring pixels denote edges. The possible edge points are grouped to see if they form edges and edges are then combined to detect object boundaries. The methods used for edge detection are known as edge-detectors or edge-detection techniques.

Figure 5.6 shows the transition of pixel intensities represented as edges. In the step edge, the pixel intensity changes abruptly, causing a discontinuity. In line edges, the value changes and then returns to the original value.

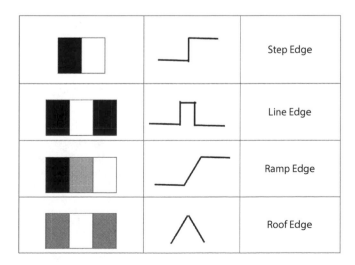

FIGURE 5.6
Different types of edges in a digital image.

For ramp and roof edges, the discontinuity occurs over a period. In practice, due to noise and other distortions, step and line edges may become ramp and roof edges.

Edges in an image are detected using first-order derivatives and second-order derivatives. Let us consider a one-dimensional function $f(x)$. The first derivative $f'(x)$ of this function is obtained as the difference between two adjacent pixels:

$$\partial f / \partial x = f'(x) = f(x+1) - f(x)$$

$\partial f / \partial x$ represents the amount of change in $f(x)$, when x changes. The second-order derivative $f''(x)$ is given by

$$\partial^2 f / \partial x^2 = f''(x) = f'(x+1) - f'(x)$$
$$= (x+1+1) - f(x+1) - f'(x)$$

Substituting the value of $f'(x)$

$$\partial^2 f / \partial x^2 = (x+2) - f(x+1) - f(x+1) + f(x)$$
$$= (x+2) - 2f(x+1) + f(x)$$

The first derivative tells us whether the function is increasing or decreasing at x. The second derivative tells us if the first derivative is increasing or decreasing. If the second derivative is positive, then the first derivative is increasing and vice versa. Figure 5.7 shows the first and second derivatives for sample images showing the transition from black to white and vice versa.

The first and second derivatives can therefore be used to determine variations (or discontinuities). Points that lie on an edge can be ascertained by

- Detecting the local maxima or local minima of the first derivative
- Detecting zero crossings of the second derivative

In other words, at constant intensity (non-edge) points, the second derivate would be zero. Edges such as step and ramp edges have changes in intensity points and hence must be nonzero at the onset of the edge. Also, it must be nonzero at the points along the intensity ramp.

5.2.2.1 First-Order Derivative Edge Detection

An image is represented by a two-dimensional function $f(x, y)$. The one-dimensional derivative explained above will be the partial derivatives applied in both spatial directions. The gradient in an image is

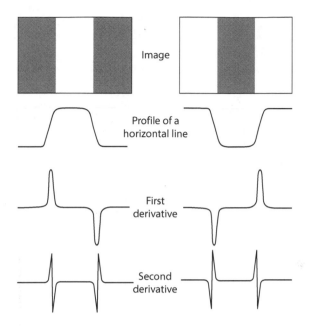

FIGURE 5.7
First- and second-order derivatives.

therefore obtained by applying the partial derivatives $\partial f/\partial x$ and $\partial f/\partial y$ at every pixel position and is given as follows:

$$G\big(f(x,y)\big)=\begin{bmatrix} Gx \\ Gy \end{bmatrix}=\begin{bmatrix} df / dx \\ df / dy \end{bmatrix}$$

The gradient magnitude is given by

$$|G|=\sqrt{Gx^2+Gy^2}$$

The gradient is commonly approximated to

$$|G|=|Gx|+|Gy|$$

The angle of orientation of the edge given by

$$\theta=\tan^{-1}\left[\frac{Gy}{Gx}\right]$$

The angle of orientation gives the direction of the largest possible increase from gradient of the image intensity at each point. For instance, it can give the direction of the largest possible increase from light to dark and the rate of change in that direction. The derivative can be implemented throughout the image using different masks and the procedure of convolution explained earlier.

5.2.2.1.1 Roberts Operator

Also known as Roberts cross or crossover edge detector, it is one of the simplest operatives used to find the edges in a given image. A 2D spatial gradient is computed for the image. Pixel points in the output image represent the estimated spatial gradient of the input image at the points. A pair of convolution masks are used to compute the first-order derivative. The second mask is the first mask rotated by 90°. A set of masks that can be used for convolution is given in Figure 5.8.

From the masks it can be seen that Gx and Gy are approximated as follows:

$$Gx = f(i,j) - f(i+1, j+1)$$

$$Gy = f(i+1, j) - f(i, j+1)$$

The kernels can be applied separately to the input image to produce separate measurements of the gradient component in each orientation. These can then be combined to determine the absolute magnitude of the gradient at each point and the angle of orientation of the edge. Note that the differences are computed at the interpolated point and not at point (i, j). Hence, this is an approximation to the continuous gradient at that point.

5.2.2.1.2 Prewitt Operator

The Prewitt operator is very similar to the Roberts cross operator except it uses 3 × 3 masks to find edges. A set of masks for Gx and Gy are given in Figure 5.9.

These convolution kernels are designed to respond maximally to edges running vertically and horizontally relative to the pixel grid, one kernel for

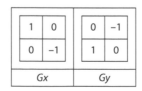

FIGURE 5.8
Roberts crossover operator.

-1	-1	-1
0	0	0
1	1	1

-1	0	1
-1	0	1
-1	0	1

Gx   Gy

FIGURE 5.9
Prewitt operator.

each of the two perpendicular orientations. Like Roberts operator, the kernels can be applied separately to the input image and combined together to find the absolute magnitude of the gradient at each point and its orientation.

5.2.2.1.3 Sobel Operator

This operator is a modification of the Prewitt operator obtained by changing the center coefficients to 2. These masks, shown in Figure 5.10, are designed to respond maximally to edges running vertically and horizontally relative to the pixel grid, one mask for each of the two perpendicular orientations.

The Sobel operator is slower to compute than the Roberts cross operator as its convolution kernel is larger. However, the operator is less sensitive to noise. The edges in the output image are made wider because of the smoothing effect of the Sobel operator.

5.2.2.2 Second-Order Derivative Operators

The edge detectors discussed in the previous section compute the edge using first-order derivative. In this method, there will be a peak in the first derivative and there would be zero crossing in the second derivative. Two operators that use the second-order derivative methods are discussed here. They are known as Laplacian of Gaussian operator and Canny edge detector.

-1	-2	-1
0	0	0
1	2	1

-1	0	1
-2	0	2
-1	0	1

FIGURE 5.10
Sobel operator.

5.2.2.2.1 *Laplacian of Gaussian Operator*

The Laplacian of an image highlights regions of rapid intensity change and is therefore often used for edge detection. The Laplacian $L(x, y)$ of an image with pixel intensity values $I(x, y)$ is given by

$$L(x,y) = (\partial^2 I / \partial x^2) + (\partial^2 I / \partial y^2)$$

Edge points detected by finding the zero crossings of the second derivative of the image intensity are particularly sensitive to noise. Therefore, it is desirable to filter out the noise before edge enhancement. The Gaussian smoothing filter is first applied to the image to reduce noise and then the Laplacian is applied. The combined operator is called "Laplacian of Gaussian."

The zero crossing detector looks for places in the Laplacian of an image where the value of the Laplacian passes through zero, that is, points where the Laplacian changes sign. Zero crossings always lie on closed contours and so the output from the zero crossing detectors is usually a binary image with single pixel thickness lines showing the positions of the zero crossing points. The convolution kernels that can be used for approximation of second derivatives are given in Figure 5.11.

5.2.2.2.2 *Canny Edge Operator*

The Canny operator works in a multistage process as follows:

1. Suppresses noise
2. Computes magnitude and direction using first-order derivative
3. Applies non-maxima suppression
4. Uses hysteresis to track the edges

1	1	1
1	-8	1
1	1	1

-1	2	-1
2	-4	2
-1	2	-1

0	1	0
1	-4	1
0	1	0

FIGURE 5.11
Kernel used by Laplacian operator.

In the first step, the image is smoothed using a filter such as Gaussian filter. Then a Roberts or Sobel operator is used to obtain the magnitude and direction. We know that edge occurs when we get a maxima. The edge obtained is still blurred and non-maxima suppression is applied to thin the edges. The edge strength of the current pixel is compared with the pixel strength of its neighboring pixels in the positive and negative gradient directions. If the pixel value is higher than the values of the neighboring pixels, the value is retained; otherwise it is suppressed.

In the next step, two thresholds, $T1$ and $T2$, are chosen, where $T1 > T2$. Tracking can only begin at a point on a ridge higher than $T1$. Tracking then continues in both directions out from that point until the height of the ridge falls below $T2$. This hysteresis helps to ensure that noisy edges are not broken up into multiple edge fragments.

5.2.2.3 Comparison of Edge Detection Techniques

Figure 5.12 compares the results of the edge detection methods discussed above. Note that sharp edges are obtained using Canny edge detectors. It can also be seen that reliable results are obtained using second-order derivatives. However, it should be noted that no single operator can fit all images. The computational complexity increases with the size of the operator. The edges obtained may also not be continuous.

5.2.3 Region-Based Segmentation

This method works on the principle of homogeneity, that is, pixels within a region have similar characteristics compared to pixels outside, which are dissimilar. The objective of region-based segmentation is to divide an image into a few large regions that are meaningful.

Region-based methods can basically be divided into

- Region growing methods
- Region split and merge methods

Good segmentation depends on the homogeneity criteria. Gray tones or textures are commonly used homogeneity criteria. Each region that is formed must be distinct and the boundary of each segment should be simple and not ragged.

5.2.3.1 Region Growing Methods

The simplest approach is to check a pixel and its neighbors for similarity (gray level, texture, color, shape, etc.). Every pixel is compared with its neighbors and those that are similar are added together to form a region.

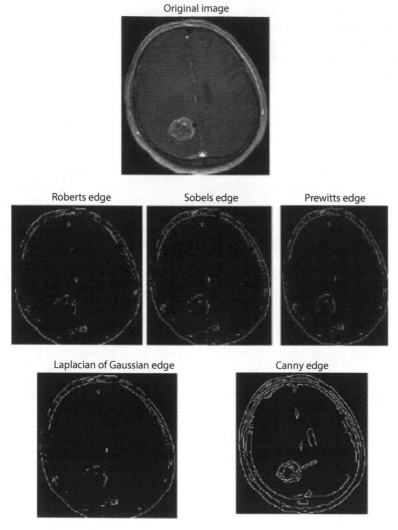

FIGURE 5.12
Results of various edge detection methodologies.

The regions are "grown" in this manner. The growing is stopped when the similarity test fails. The selection of similarly criteria is therefore significant.

Region growing method is found to give reliable results. Region growing methods can be further classified as seeded region growing (SRG) and unseeded region growing (UsRG) methods. The main difference between SRG and UsRG is that SRG is semiautomatic in nature, whereas UsRG is fully automatic in nature.

5.2.3.1.1 Seeded Region Growing Method

In this approach, the seeds are initially specified by the user. A seed is a test pixel with ideal characteristics that belongs to the region of interest in the subject image. The choice of seed is crucial since the overall success of the segmentation is dependent on the seed input. The seed set can have one or more members as per the user's choice. For example, for the bowl of fruit shown in Figure 5.5, if the objective is to detect the bananas in the plate, the seed characteristics would match the banana. Likewise, separate seeds could be specified for each type of fruit in the bowl.

For the given set of seeds, pixels are added to one or more of the seed sets. The process is semiautomatic in nature as the user specifies the seed, while the growth process is automatic. There are several algorithms available for segmentation using the SRG approaches. While the overall methodology remains the same, the difference lies in the approach to comparing pixels for adding to the regions, also called as clusters. The general steps for SRG methodology are as follows:

1. Determine seeds to start the segmentation process.
2. Determine the similarly criteria for each region. As mentioned earlier, there can be multiple regions and there should be no ambiguity in the selection criteria that specifies the criteria for adding pixels to a particular region.
3. All pixels in the image are checked and assigned to different regions.
4. If two regions are found to have similar characteristics, they can be merged to form a single region.

Figure 5.13 illustrates the seeded region growing methods with an example. From the Figure, it is evident that regions are grown through multiple iterations. Note that the regions are clearly defined by the 90th iteration. The white circle in the figure indicates the inconsistent edge segments.

SRG has been found to be robust, quick, and free of tuning parameters, which makes it suitable for a large range of images. The main drawbacks of this method are twofold. The initial seed points are decided by the user. The segmentation results depend on the seed points chosen and can be different for different sets of seed points. This problem, therefore, reduces the stability of the segmentation results. Also, there is the important issue of how many seed points are chosen, as it affects how the image is segmented. The second more serious drawback is that SRG requires lots of computation time.

5.2.3.1.2 Unseeded Region Growing Method

Unseeded region growing methods are flexible and fully automatic. Seeds need not be explicitly specified by the user, but are generated automatically by the process. The general approach is as follows:

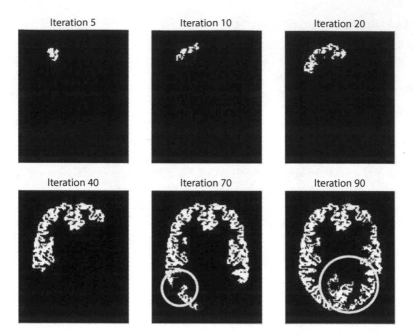

FIGURE 5.13
Example of seeded region growing method.

1. The segmentation process is initialized with a region A1, starting with a single pixel that eventually results in several regions (A1, A2, A3, ... An) on completion.

2. The test pixel is compared with the pixel belonging to the region A1. If it satisfies the selection criteria, then the pixel is added to the region A1, otherwise a new region is formed. In this manner, each pixel is matched with pixels of the regions already defined and added to the region it matches. If it does not match the characteristics of any of the already defined regions, then a new region is formed.

3. After a pixel is allocated to a region, the mean pixel value of the region is recomputed.

4. The steps are iterated until all pixels are assigned to a region.

Figure 5.14 illustrates the method with an example. The problem of inconsistent edge segments obtained with seeded methods can be avoided with this method as indicated by the white circle in the Figure 5.14.

5.2.3.2 Region Split and Merge Method

Like the previous region segmentation methods, homogeneity criterion is used to split the image into regions. The image is successively split into

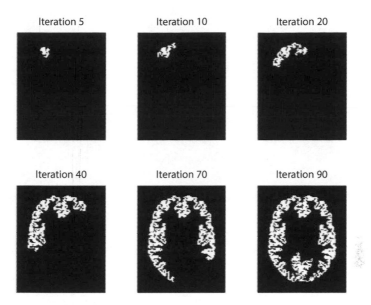

FIGURE 5.14
Example of unseeded region growing method.

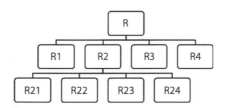

FIGURE 5.15
Region split and merge example.

quadrants based on the specified homogeneity criteria. Similar regions are merged together to create a segment.

Figure 5.15 shows an image represented by R. Quad trees are used to implement this method. In quad trees, each node of a tree has exactly four children. The root of the tree represents the entire image. In the example, R has four child nodes: R1, R2, R3, and R4. R2 is further split into four descendants: R21, R22, R23, and R24.

For the image R, let the homogeneity criteria be defined as predicate P. Then the steps are as follows:

1. Split the image R into four equal regions.
2. Apply P to a region. If P(R) is false, then split the region into four quadrants.

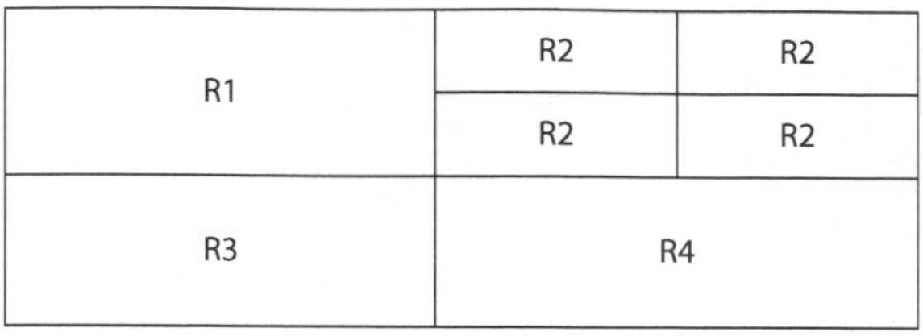

R1	R2	R2
	R2	R2
R3	R4	

FIGURE 5.16
Region splitting process.

3. Continue this process until P(R) is true.

4. Merge regions where the homogeneity condition matches.

5. Repeat for all regions.

Figure 5.16 illustrates the region splitting process.

Homogeneity can be defined in different ways. It can be defined using threshold as explained earlier. The region is said to be homogenous if it lies within specific thresholds. Mean and variance of intensity levels can also be used to check homogeneity.

5.3 Object Recognition

An object has to be detected before it can be recognized. The goal of detection is to distinguish objects from their background. Typically, objects may have to be detected against cluttered, noisy backgrounds and other objects under different illumination and contrast environments. Object recognition attempts to identify the detected object by comparing with a known image of the object. Before proceeding with the discussion of the different methods and techniques, let us understand the various factors that contribute to making object recognition difficult:

1. **Lighting**: We know that proper lighting is important to get good quality images. In practice, the lighting conditions may vary during the day. Also, the weather conditions may affect the lighting in an image. Indoor and outdoor images of the same object can look quite different due to varying lightning conditions. Shadows in the image can affect the image light.

2. **Positioning**: The position of the object may not be correct. For example, when an object moves on the conveyor, its orientation and position may change due to vibration. Hence, the image obtained may

not be of good quality. Also, the image captured may be only a partial image and not the full image. So image recognition would become difficult.

3. **Rotation**: The character or letters in the image may get rotated because the orientation of the object is different. Hence, in the image captured, it would be difficult to recognize the object. For example, if the characters of a label get rotated in the image, recognition of characters would be difficult.

4. **Mirroring**: The mirrored image of any object is difficult to recognize.

5. **Occlusion**: The condition when object in an image is not completely visible is referred as occlusion. Occlusion can happen because of shadows caused by lighting and also because one object hides another object. So only a partial object may be seen in the image.

6. **Scale**: A change in the size of an object in the image as compared to the original object size would make comparison difficult.

Proper feature representation of an object is a crucial step in an object recognition system as it helps to identify and distinguish the object from the background or other objects in different lightings or scenarios. Object recognition features are categorized in two ways—sparse and dense representations. For sparse feature representations, interest point detectors are used to identify structures such as corners and blobs on the object. Dense representation is used to identify complex structures such as irregular shapes.

There are many object recognition methods and techniques, and many algorithms have been proposed that use these techniques. The majority of these methods produce reliable results under constrained scenarios. Basic assumptions are made to reduce the number of complicating factors that are inherent in object detection and recognition. Common assumptions concern object appearance, background color intensity information, duration of time for which an object exists in the scene, objects occlusion, viewpoint invariance, illumination, limitation regarding number of objects within the scene, among others. However, in practical applications it may not always be possible to obtain images that satisfy these assumptions. We will discuss two of the common matching techniques:

- Template matching
- Blob analysis

5.3.1 Template Matching

Template matching is a method that finds objects by pixel matching. A template of the object to the recognized is created. A template is an image of the object taken under standard conditions. The template could be a sub-image that is matched against the whole image. For example, going back to our bowl of fruits in Figure 5.5, a sub-image or template could be that of an apple which is matched with the full image to find the presence of apple fruit.

In this technique, a pixel-by-pixel matching is done between the image where object has to be recognized and the template. The template is placed at every possible pixel of the image. The problem is to find a similar occurrence in the target image. Again, the similarity may not exist if the pattern is not present in the target image. For example, template matching is used to identify characters, numbers, and other similar objects. It can also be used to detect edges of figures, navigate robots, etc.

Figure 5.17 shows an example of template matching. There are two images in the figure. The first is the template and the second is the target image in which objects are to be recognized. This example is used to recognize the letter "N." There are two occurrences of "N" in the target image in the word "MACHINE VISION." Noted that the template or pattern is usually smaller than the image with which it is compared.

Matching is done using similarity measures. One of the similarity measures that is used for matching is the correlation measure. Cross-correlation between the two images is computed as follows:

$$\text{Cross-correlation}(\text{image1}, \text{image2}) = \sum_{x,y} \text{image1}(x,y) \times \text{image2}(x,y)$$

The similarity measure is computed between the template and target image at point (x, y) in the target image. The cross-correlation is essentially a simple sum of the pair-wise multiplication of the template and the target image. The value obtained would be maximum for locations where there is correspondence. Cross-correlation is susceptible to global brightness of the image. Hence, if an image is brightened, match may occur where none exists. Normalized cross-correlation (NCC) gives a better similarity measure and is given as follows:

$$\text{NCC}(\text{image1}, \text{image2})$$

$$= 1/N\sigma_1\sigma_2 \sum_{x,y} \left(\text{image1}(x,y) - \overline{\text{image1}}\right) \times \left(\text{image2}(x,y) - \overline{\text{image2}}\right)$$

N		MACHINE VISION
Template		Target Image

FIGURE 5.17
Template matching example.

From the equation, it is clear that the images are normalized by subtracting the mean pixel value and dividing by the standard deviation. The maximum values occur for location where the sub-image of $f(x, y)$ perfectly matches the template. Other techniques such as sum of squared difference or sum of absolute differences, pyramid modeling, and multiresolution analysis can also be used for matching.

Template matching is done as follows:

1. The target image can be in any one of the file formats, such as JPG/JPEG, PNG, etc.
2. The target image is first converted into binary image or a grayscale image. This is done to reduce computation time and complexity.
3. The target image is matched with the template.

Template matching methods can vary in their approach to matching. In feature-based approaches, image features, such as shapes, textures, colors, and edges, are extracted and matched with the template. In model-based approaches, boundaries of objects are detected in the target image and compared with the template. Area-based methods combine feature and model matching. Different similarity measures used for matching are shown in Figure 5.18.

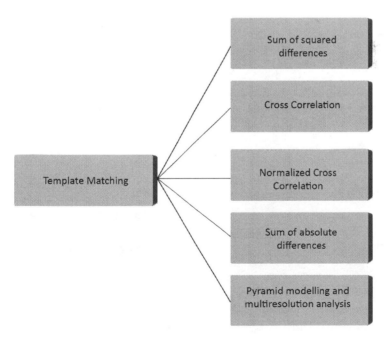

FIGURE 5.18
Different methods of template matching.

5.3.2 Blob Analysis

Blob stands for "binary large object." The method of analyzing an image which has undergone binarization processing is called "blob analysis." Blob analysis is one of the basic methods used for analyzing the shape of an object. In blob analysis, we first separate the different objects in an image and then try to evaluate which object we are looking to recognize. For example, the objective may be to look for circles, squares, or other shapes present in a target image.

A group of connected pixels is called a blob, which is also referred to as "lump." Connection between two pixels is defined by connectivity, that is, which pixels are neighbors and which are not. The two most often applied types of connectivity are 8-connectivity and 4-connectivity; 8-connectivity is more accurate than 4-connectivity, but 4-connectivity is often applied because it requires fewer computations and so image processing is faster.

The steps involved in blob analysis can be enumerated as follows:

1. **Extraction**: Threshold-based segmentation is applied to obtain the regions in the image.
2. **Refinement**: The extracted region contains noise due to reasons of inconsistent lightning and poor image quality. The image is refined using techniques such as interpolation to smooth the boundaries.
3. **Analysis**: Each of the regions is analyzed to determine the blobs that match the template. The regions may contain one or more objects that can be split as individual blobs for matching.

Figure 5.19 illustrate the principle of blob analysis for a multi-pin integrated circuit (IC). The number of pins present in the input image is calculated by setting up a region as a blob, shown using a white rectangle at the top of the figure. The respective width and height of the pin are measured by setting the blob vertically, shown using a white rectangle in the figure.

5.4 Summary

Digital image processing has been discussed under the broad headings of preprocessing, image segmentation, and object recognition. Preprocessing is carried out to remove noise and improve the quality of images. Different types of filters are used to remove the noise in images. Another important preprocessing step is scaling, which is done to scale up or scale down images to the required size. Frequency analysis is carried out using histograms. Brightness and contrast of images can be adjusted to improve their quality. Different methods for image segmentation such as threshold-based, edge-based, and region-based have been discussed, as were object recognition methods such as template matching and blob analysis.

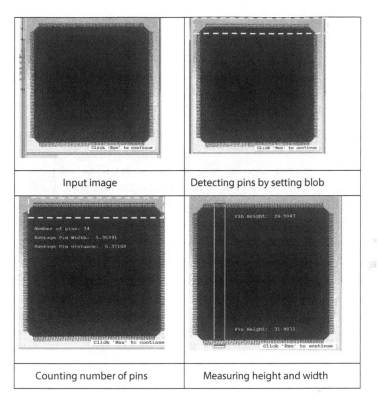

Input image	Detecting pins by setting blob
Counting number of pins	Measuring height and width

FIGURE 5.19
Sample images illustrating blob analysis.

The methodologies that need to be adopted at every step depend on the specific application requirements. While the list of methods is endless, only a few typical and commonly used methods were discussed in this chapter. The References section includes many references for image processing and it is recommended that Readers use these references for additional reading and understanding.

Exercises

(Hint: You can use any image processing software to obtain solutions.)

1. Read a grayscale image and express it as an array or image matrix.
2. Create a 5 × 5 array and convert it into an image. Convert the obtained image into a binary image.

3. Read a color image and convert the image first to a grayscale and then to a binary image.

4. Obtain the inverse of the following types of images:
   a. Binary image
   b. Grayscale image
   c. Color image

5. Read any grayscale image "x.bmp." Display the image and its histogram. Then perform histogram equalization and display the image obtained after the equalization. Compare the two images and explain the difference. What happens if histogram equalization is done twice?

6. Obtain the histograms for a color image.

7. Take any two grayscale images; perform the operations of ADD, SUBTRACT, MULTIPLY, and DIVIDE; and display the resulting images.

8. Overlay a small image onto a larger image and display the combined image.

9. Explain thresholding. Where can it be used in image processing?

10. For any image X, perform the different types of filtering and compare the images obtained. Explain the difference in the images obtained.

11. Explain the difference between object detection and object recognition using examples.

12. What is object classification? Where is it used in image processing?

13. What is the difference between image segmentation and edge detection?

14. Explain how the following similarity measures are calculated:
    a. Sum of squared difference
    b. Sum of absolute difference

15. Differentiate between edges and contours in an image.

16. Apply seeded and unseeded region-based segmentation on an image. Discuss the results obtained.

17. What are the different types of methods used in template matching?

18. Find the circles in an image that has different geometric shapes using blob analysis.

19. What is multiresolution analysis in image processing? Explain.

20. Explain any two object recognition techniques used for face recognition.

6

Case Studies

Designing a machine vision solution is a challenging task. Each application is unique and comes with its own requirements and desired outcomes. The required results need to be achieved at optimum cost while efficiently meeting performance requirements. In this chapter, we use case studies to explain the design process. In this chapter, we use case studies from quality inspection domain to illustrate the design process. The case studies provide a general approach to vision system design and selection of machine vision components like camera, lens, and lighting.

It is needless to add that the design of the machine vision system is influenced by many factors. The part to be inspected is obviously an important criterion. Factors such as the size of the part or defect, the texture of the part, the type of inspection required, and the speed required for performing the task all influence the design. The environment of the factory, particularly at the inspection area, will also play an important role in the design of the system. Factors such as temperature, humidity, and vibration may influence the choice of machine vision components, which may also need protection to withstand the rigors of harsh conditions.

How the quality inspection system interfaces with the existing production process, as well as whether interaction with a human operator is required, will also impact the design of the vision system. The information or results that are required is another important criterion. For example, it may be required to provide a visual display unit to show the results to a human operator/supervisor. It may also be needed to log all data pertaining to the quality process for tracking and analysis.

6.1 Case Study—Presence/Absence Inspection of a 3G Switch Box

This case study illustrates how to check for presence/absence of components in a part. The part to be inspected is an electronic switch box for 3G mobile communication, which is shown in Figure 6.1. The box is the final product that is to be inspected prior to packaging and dispatch.

FIGURE 6.1
3G switch box (front and back views)—case study

6.1.1 Inspection Requirements

The 3G box has connectors at the front, while the fans, warning label, and barcode of the product are located at the back. Let us assume that the quality inspection needs to check for the following:

1. Front of the box
   a. Presence/absence of connector covers
2. Back of the box
   a. Presence/absence of fan
   b. Presence/absence of warning label
   c. Presence/absence of barcode

The box is to be inspected as it moves along on an automated belt. Let the length of the box be 15 mm. While designing the vision system, it is assumed that the parts are presented individually without overlap. The automated belt will be stopped for 10 seconds for capturing the image for inspection. The total time allowed for processing a box is 20 seconds. During this period, a decision has to be made as to whether the box is OK or NOT OK. If the box does not pass the quality inspection, it has to be rejected. The objective of the inspection is to check for the presence or absence of components in the box. Edge detection would be an appropriate technique to carry out the image analysis.

6.1.2 Machine Vision Configuration

In this example, we have assumed that the switch box will be static for 10 seconds for inspection. Hence, an area scan camera is selected to capture the image in the 10-second period. Area scan cameras are easy to set up and operate, so they are used extensively in machine vision applications.

To decide on the camera resolution, we need to compute the Field of View (FOV). The length of the switch box is 15 mm, but it may not be possible to always position the box exactly in the required position for the camera. Vibration of the belt may cause the part to move from its intended position. Let us assume a tolerance of ±1 mm in positioning the part for image capture. Let us further assume a margin of 1mm in image capture to provide for space between the part edge and the image edge.

Next, we need to choose a camera with the appropriate aspect ratio. Aspect ratio is the relationship between the width and height of an image. Aspect ratio is given by the size of the camera sensor, taken from the width and height of image. For example, if the camera sensor is 36 mm wide and 24 mm high, its aspect ratio would be 3:2. Let us assume an aspect ratio of 4:3 as most compact cameras come in this size. FOV is calculated in the vertical and horizontal directions as follows:

$$FOV_{Ver} = \text{Maximum part size} + \text{Tolerance in positioning} + \text{Margin}$$

$$= 15 + 1 + 1$$

$$= 17 \text{ mm}$$

$$FOV_{hor} = FOV_{ver} \times \frac{4}{3}$$

$$= 17 \times 4/3$$

$$= 22.66 \text{ mm}$$

The field of view is therefore 17 × 22.66 mm.

Next, we can calculate the resolution of the camera. We know that the camera resolution is given in pixels. We need to arrive at a mapping of the part size in millimeters to the resolution R given in pixels. Spatial resolution R_s refers to clarity of an image and is defined as the smallest visible detail in an image. For our case study, we define spatial resolution as number of pixel values per millimeter. In other words, it is the number of pixels mapped to the smallest size that is to be measured. Let the smallest detail S_d be measured to be 0.1 mm for our case study. This is largely based on the measurement accuracy that is required. Let us map 1 pixel for the smallest detail. Then,

$$R_s = \frac{\text{Smallest detail to be measured}}{\text{Number of pixels to map the smallest details}}$$

$$= S_d/N_d$$

$$= 0.1/1$$

The resolution of the area scan camera calculated in both the horizontal and vertical directions is

$$R_{hor} = \frac{FOV_{hor}}{R_s}$$

$$= 22.66/(0.1/1)$$

$$= 226 \text{ pixels}$$

$$R_{ver} = \frac{FOV_{ver}}{R_s}$$

$$= 17/(0.1/1)$$

$$= 170 \text{ pixels}$$

A camera with a resolution of 480 × 240 pixels can satisfy the resolution requirement. A standard 1 megapixel (MP) camera can be chosen for our application.

The inspection needs to be carried out at the front and rear of the 3G box. We can, therefore, decide on using three cameras for image capture. One camera at the front and another one at the back. An additional camera can be placed at the front for complete coverage of the component.

Horizontal field of view can be used to decide the focal length. Since, the inspection is to check for presence or absence of detail in the part, a normal lens of focal length greater than 22.6 mm obtained through the FOV calculation can be used. Lens size is of 5, 13, 15, 17, 20, 25 mm, etc. Since the nearest higher lens size is 25 mm, it is better to opt for a standard lens size of 25 mm. Regarding lighting, for the given inspection requirements, front lighting would be a suitable choice.

What we have discussed so far is a possible choice for the camera and lighting based on the assumptions that we had made. As mentioned earlier, there are several factors that mandate the choice of machine vision equipment. For example, if a higher resolution is required to obtain more detail of the part to be inspected, then we can choose a camera with higher pixel rating, that is, 1.3MP, 2MP, 3MP, or higher. Similarly, choice of lighting is influenced by the texture of the part and the features of the part that have to be enhanced or emphasized for inspection. The ambient lighting and the environment dictate the choice of machine vision equipment in terms of how rugged and robust they must be. Availability and cost are important practical considerations.

Proceeding further with this case study, we have to decide on the power requirement. Power supply of 24V should be sufficient to support the compact cameras and front lighting. The cameras and light domes can be mounted on aluminum stands or profiles for easy adjustment. Aluminum is light weight and can withstand wear and tear.

Many image processing software packages are available for use in machine vision systems. They provide powerful image processing functionality that can be used to develop the software.

6.1.3 Machine Vision Setup

The machine vision setup is shown in Figure 6.2. It can be noted that the 3G box to be inspected will arrive from the left-hand side (LHS). The box to be inspected can be placed manually by the operator or it can be made to move automatically after production. The belt or conveyor will carry the box into the closed inspection area. Sensing the placement or arrival of the box, the light will flash, and the cameras will capture the images on the front and back ends. These images will be displayed on the monitor in the control panel. The captured image is processed to determine the presence/absence of the connector covers on the front, and the fan, barcode, and warning label on the back. A green light on the control panel will be lit to indicate the box is OK and the box will be moved to the right-hand side (RHS) and from there on to packing and dispatch. If the box fails to pass the inspection check, a red light will be turned on and additionally a buzzer alarm can be sounded.

FIGURE 6.2
Vision setup for switch box inspection case study.

6.2 Case Study—Surface Inspection of a Rivet

This case study illustrates the surface inspection of a part. The part to be inspected is an aluminum rivet shown in Figure 6.3. The rivet is used to join mechanical parts to form a larger machine part. A rivet consists of a smooth cylindrical shaft with a head on one end. The end opposite to the head is called the tail. During installation, the rivet is placed in a punched or drilled hole and hammered in. Inspection of the rivet's appearance can give a good indication of the quality of the part. In this case study, the surface smoothness is checked to verify if there are burrs or scratches present on the surface. Hence, this task can be categorized as surface inspection.

6.2.1 Inspection Requirements

Let the size of the rivet be 5 mm. Let the surface smoothness of the rivet be measured with an accuracy of 0.1 mm. Let the time frame for processing be 2.5 seconds. The requirement in this case study is to inspect the surface for smoothness. Surface discontinuities like burrs or scratches should not be present. Blob analysis can be chosen as an appropriate technique for image analysis.

6.2.2 Machine Vision Configuration

In this case study, it is assumed that the part positioning of rivet is indexed by the use of an automated belt and all the parts are presented without overlap. The tolerance of part positioning is less than ±1 mm across the optical axis and ±0.1 mm in the direction of the optical axis. The belt stops for 2 seconds. The maximum part size as stated earlier is 5 mm. As the requirement is for surface inspection of a single object as it moves on the automated belt, an area scan camera can be used to capture the image.

Following the steps outlined in the previous case study, FOV and camera resolution can be calculated. For this case study, we assume the aspect ratio is 5:3, which is another standard aspect ratio in cameras. FOV is calculated in the vertical and horizontal directions as follows:

FIGURE 6.3
Aluminum rivet—case study.

FOV_{Ver} = Maximum part size + Tolerance in positioning + Margin

In this example, based on the requirements and specification,

maximum part size = 5 mm
tolerance in positioning = 1 mm
margin = 1 mm
aspect ratio = 5:3

Hence, the field of view is

$$\text{FOV}_{\text{ver}} = 5\,\text{mm} + 1\,\text{mm} + 1\,\text{mm}$$

$$= 7\,\text{mm}$$

As the aspect ratio of the camera sensor is 5:3, the FOV in the horizontal direction is adapted to

$$\text{FOV}_{\text{hor}} = \text{FOV}_{\text{ver}} \times \frac{5}{3}$$

$$= 7\,\text{mm} \times 5/3$$

$$= 11.66\,\text{mm}$$

As the FOV and the accuracy of the measurement are known, the necessary sensor resolution can be calculated as before:

$$R_c = \frac{\text{FOV}}{R_s}$$

$$= \text{FOV} \times \frac{N_d}{S_d}$$

The requirement specification is for an accuracy of 0.1 mm and we will map 1 pixel for the smallest detail. The horizontal and vertical resolutions can be evaluated as:

$$R_{\text{chor}} = 7\,\text{mm} \times \frac{1\,\text{pixel}}{0.1\,\text{mm}} = 70\,\text{pixels}$$

$$R_{\text{cver}} = 11.66\,\text{mm} \times \frac{1\,\text{pixel}}{0.1\,\text{mm}} = 116.6\,\text{pixels}$$

1MP camera would be able to provide the required resolution. Further, one camera should suffice for inspecting the surface smoothness.

We can choose a telecentric lens for surface inspection. A telecentric lens (refer to Chapter 3) has many advantages. When the part moves through the FOV, the positioning of the part will vary due to the vibration of the belt. Consequently, the working distance for image capture will change for each part. In telecentric lenses the magnification does not change when working distance changes. Hence, we can get the high accuracy required for imaging the part.

The FOV was computed as 11.66 mm in the horizontal direction; hence, a lens with a FOV of 12.5 mm can be chosen. The working distance between the camera and the object can be adjusted to obtain an image of good clarity.

The next step is to choose the lighting. We want to enhance the textural or surface features of the part for inspection. Hence, a dome light can be chosen as it diffuses front light. LED light source can be chosen for its longevity.

6.2.3 Machine Vision Setup

The setup of a camera, its illumination, and the part are displayed in Figure 6.4.

No special mechanical support or design may be required. Cameras and light domes can be mounted on aluminum profiles for easy adjustment. 24 V power supply should be sufficient to support the power requirements. The software can be developed using built-in libraries provided by image processing software.

The rivets will move on the conveyor belt and be inspected as they come within the frame of the machine vision setup. On sensing a rivet, the light will flash, and the image will be captured by the camera. A control panel can be used to display the image and the results. The rivets can be segregated as OK or NOT OK into separate bins using automation setup.

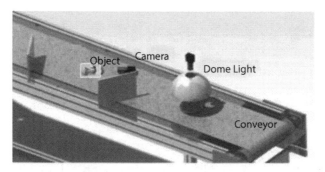

FIGURE 6.4
Machine vision setup for rivet surface inspection case study.

6.3 Case Study—Dimensional Measurement of a Cage Sleeve

This case study has been chosen to illustrate dimensional measurement. The part to be inspected is the cage sleeve shown in Figure 6.5. This case study also discusses the use of line scan cameras for image capture.

6.3.1 Inspection Requirements

In this case study, we need to measure the dimension of the top opening as well as the dimensions of the six openings on the side of the component. These measurements have to be carried out using an accuracy of 18 microns and the inspection parameters as seen from the figure are as follows:

1. Window width and height
2. Pillar width and height for the side openings

Let the window width be 7 mm and the height be 3 mm. Let the pillar height be 6.5 mm and width 3.2 mm.

6.3.2 Machine Vision Configuration

For this case study, we will use line scan cameras to continuously capture images from the stations as the cage sleeve parts move along on a conveyor belt. We know that line scan cameras are best suited for applications where there is a need for large or high-resolution or high-speed capture.

Two camera setups along with appropriate lighting are needed: one setup (or station) to measure the top opening and one to measure the six openings on the side. A line scan camera captures one image line at a time. So, to capture the 2D image of the object, either the camera or the object needs to be rotated. We will mount the camera and lighting on a base and rotate the base using

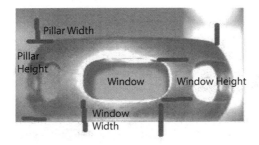

FIGURE 6.5
Sample cage sleeve—case study.

a servo system to image all six openings on the side. It is important that the camera movement is synchronized with the movement of the conveyor, so the motion is uniform, and the images are obtained correctly. The speed can be synchronized using encoders.

A TDI line scan camera can be chosen for image capture. As explained in Chapter 3, this sensor technology involves image data in one sensor being accumulated with several other lines, synchronized with the movement of the part.

For this case study, we will determine the camera resolution using FOV and number of pixels required to cover the minimum defect size. We are going to choose a camera that has enough resolution to have at least 3 pixels for minimum defect size. For the maximum measurement of 7 mm, let us assume a minimum defect size of 0.005 mm. Tolerance and margin will be fixed as per the previous case studies. Let the aspect ratio of the camera be 4:3. FOV is calculated as follows:

FOV_{Ver} = Maximum part size + Tolerance in positioning + Margin

FOV_{ver} = 7 mm + 1 mm + 1 mm = 9 mm

As the aspect ratio of the camera sensor is 4:3, the FOV in the horizontal direction is adapted to

$$FOV_{hor} = FOV_{ver} \times 4/3$$
$$= 9 \text{ mm} \times 4/3$$
$$= 12 \text{ mm}$$

Since the camera is a line scan camera, to calculate resolution we use the horizontal FOV.

$$\text{Resolution} = FOV/(\text{Minimum defect size}) \times (3 \text{ pixels coverage})$$
$$= (12/0.005) \times 3$$
$$= 7,200 \text{ pixels}$$

We could use an 8K (8,192 pixel) line scan camera, or two side-by-side 4K (4,096 pixel) line scan cameras with some overlap in their horizontal fields of view. In this case study, as we need two stations, one for top measurement and the other for side measurement, we will use two 4K cameras.

6.3.3 Line Rate and Resolution

The line rate can be computed from FOV, part speed, and object pixel size. Let the part speed be 60 mm per second. We are using two 4K (4,096 sensor pixels) camera. The object pixel size would be calculated as:

$$\text{Object pixel size in the FOV} = \text{FOV}/(\text{camera size in sensor pixels})$$
$$= 12/4{,}096$$
$$= 0.00292$$
$$= 74 \text{ microns}$$

The line rate is then calculated as follows:

$$\text{Line rate} = 60/0.00292 = 20.547 \text{ kHz}$$

The lens gathers light from the object to improve sensitivity and magnifies (or minimizes) the FOV to match the size of the camera's line sensor. Hence, it is essential to compute magnification factor. For the assumed camera pixel size of 7.04 microns, magnification is calculated as:

$$\text{Magnification} = \frac{\text{Camera pixel size in microns}}{\text{FOV object pixel size}} = \frac{7.04 \text{ microns}}{74 \text{ microns}} = 0.0951$$

So, there is a reduction in size of $1/0.0951 = 10.5$ from the real-world FOV to the camera's line sensor. This ratio is called "inverse magnification." Next, we must specify working distance, that is, the distance from the camera's faceplate to the object being imaged. From this we can get the focal length of the lens required:

$$\text{Focal length} = \frac{(\text{Working distance})}{\text{Inverse magnification} + 1}$$

In this example, let the working distance be 290 mm. Then,

$$\text{Focal length} = 290/(10.5 + 1)$$
$$= 25.1 \text{ mm}$$

25.1 can be approximated to 25 mm, which is a standard focal length for lenses. The working distance is usually adjusted to match a standard lens focal length.

Smaller line scan cameras, up to 2,048 pixels in some cases, can use a C-mount lens with a 1" aperture. For longer line scan sensors an F-mount, M42, or M72 mount lens is required.

6.3.4 Machine Vision Setup

The vision setup is shown in Figure 6.6. The cameras and other vision components are located within the workstation. Parts are fed into the vision system using part feeder. It can be noted that a door is provided in the workstation for the installation and maintenance of the machine vision components.

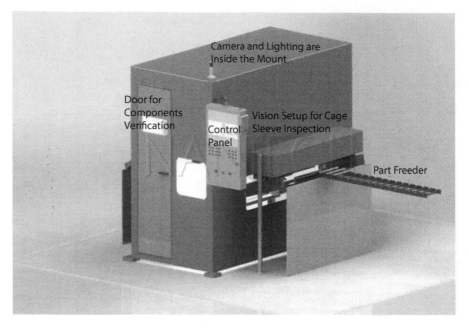

FIGURE 6.6
Vision setup for cage sleeve inspection case study.

6.4 General Process for Building Machine Vision Solutions

We know that a well-designed machine vision system can help manufacturers improve product quality and gain customer satisfaction. Hence, the starting point for deciding on a machine vision solution is the possible benefits that can be accrued by such a system. Cost is an important factor. Project costs include cost of machine vision components, software acquisition and development, automation costs, integration costs, installation cost, operator training, etc. A cost-benefit analysis can be carried out to estimate the overall cost and quantify the possible benefits of implementing the machine vision system.

Once we decide to go ahead and design a machine vision solution, the first step would be to define the requirements of the system. For a quality control system, the quality inspection requirements have to be specified in detail. In Chapter 4, we looked at different examples of quality control applications from the industry. The quality inspections include dimensional measurement, checks for presence/absence of components, label inspections, barcode verification, etc. Each of these tasks require different vision requirements.

A machine vision solution that works for one task may not be suitable for another type of task or inspection. Hence, each solution is unique and has to be designed to meet the outcomes of a specific application. Some of the questions that need to be asked to arrive at the requirements of the application include the following:

- Is the part a discrete object or is it continuous (paper or cloth)?
- Can the part be momentarily stationery for image capture, or will it be continuously moving?
- Is the shape of the object regular or irregular?
- What type of inspection needs to be carried out?
- What features of the part need to be extracted?
- What type of image processing methodology is best suited to the application?
- Is it monochrome or color image processing?
- Will the parts be separate or will they overlap?
- What are the timing requirements?
- What are the other performance requirements?

Once we know the requirements, we have to decide on the machine vision configuration that meets the application requirements. In the case studies discussed above, we have given some practical examples for machine vision component selection and have made assumptions to illustrate the design process. We know, however, that many factors have to be considered to arrive at a configuration.

Off-the-shelf components are available to build a machine vision solution. Readily available solutions may be offered by vendors with limited customization to tweak the system to suit customer requirements. This would result in a shorter deployment time. Vendors may also offer an integrated or end-to-end solution that covers the inspection tasks, automation design and implementation, installation, and ongoing support. Vendor selection is an important aspect for buying individual components as well for solutions. Vendors should be able to offer proper technical support and should also have a sustainable presence in the market.

Once the machine vision system is ready—either built or bought out—the next important step is to test the system. It is advisable to design acceptance tests to prove the vision system meets the requirements. Samples used for testing must be representative of the actual products manufactured. Feasibility testing for all types of possible defects must be carried out.

Finally, when you go live with the application, it is necessary to ensure that you have the means to debug the system immediately. If a problem occurs, a solution must be readily available. Remote access may be provided for debugging.

6.5 Summary

In this chapter, we looked at three different case studies from the quality control domain. The first case study was a 3G box inspection to check for presence/absence of items on both the front and back of the box. Since the imaging requirements were high, three cameras were used for this quality inspection. The second case study was a surface inspection of rivets. A telecentric lens was chosen to get high accuracy in imaging. The third case study was inspection of a cage sleeve and line cameras were chosen for the inspection process. For each of the case studies, the inspection requirements were explained and selection of machine vision components was discussed. A general process for building a machine vision solution has been presented.

In summary, when designing a machine vision solution, it is essential that the decision of the machine vision configuration is made, taking into consideration all technical and practical aspects, such as the inspection requirements, the part specifications (texture, size, material, etc.), required speed of processing, prevailing factory environment, and cost.

Exercises

(Hint: You can use any image processing software to obtain the solution.)

1. Write a detailed note on the steps involved in the design of a machine vision system.
2. Take a sample application and do the cost-benefit analysis of designing a machine vision system. Explain all of your assumptions.
3. Give three examples each for use of line scan cameras and area scan cameras. (Avoid examples from this book.)
4. Explain the design process for configuring a machine vision setup for doing profile inspection. State any assumptions made.
5. Explain the advantages and disadvantages of building or buying a machine vision system.
6. Do a study to compare the different imaging processing software currently available. Include both open-source and licensed software (purchased).
7. Do a study to identify the different resolutions of camera (both area scan and line scan) currently available in the market.

8. Identify inspection requirement for checking the presence and absence of components in a mobile phone. Draw a neat diagram of the mobile phone and explain in detail the inspection parameters.

9. Explain what the challenges are involved in designing a machine vision system for automated inspection of syringe needle tips. (Hint: The syringe is to be used for giving injections to human beings and therefore requires a clean and sterile environment.)

10. Design a machine vision solution to segment out the ripe crops from the image shown below.

11. Design a machine vision solution to read the barcode that is etched on the component. Use any sample object to design and test your system.

12. Develop a machine vision system to segregate multiple backgrounds present in the same scene. Test using a sample image.

13. What are hyperspectral and multispectral imaging? Compare the two types of imaging and explain the difference.

14. Design a machine vision system to measure the diameter of a rough, deep, and small (0.5 mm) hole.

15. The iris of the eye is used for identification of humans in a biometric security system. Discuss the challenges involved in designing a machine vision system for iris segmentation and normalization. Design and test your system.

16. Discuss camera selection for detecting objects in front of a moving vehicle (automobile). The camera is to be mounted on the windshield of the car.

17. Design and test a machine vision system to determine the distance between two parallel lines.

18. Design a complete vision system to inspect the following parameters of an oil seal.

a. Tear on the outer surface

b. Presence/absence of spring

c. Dimensional accuracies of inner/outer diameter

19. Design a system to recognize the characters present on the label below.

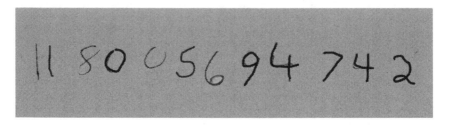

20. Design a machine vision system to recognize the different types of faces shown below.

7

Emerging Trends and Conclusion

In this last and concluding chapter, we will look at emerging trends in machine vision technology. We are currently in the era of Industry 4.0, which is witnessing the integration of many key technologies such as cyber-physical systems (CPS), artificial intelligence (AI), the Internet of things (IoT), big data analytics, cloud-computing 5G mobile communications, robotics, and many more. But before we move ahead, we need to look at the history of Industrial Revolution and see how industry has been transformed over the ages.

7.1 History of Industrial Revolution(s)

Industrial revolution has brought about remarkable changes in all social and economic aspect of human life and the way of living. While on the one hand, industrialization has led to economic prosperity, it has resulted in more population and urbanization. On a more sober note, **industrial** pollution has also led to environmental degradation and depletion of natural resources.

Let us trace the history of industrial revolutions to understand the shift from a predominantly rural agrarian society to the industrial and urban society that we are in now. Prior to the First Industrial Revolution, which took place in Britain, most people living there were farmers who resided in villages or small rural communities. The community was self-sufficient, and the main occupation was agriculture. Only a small portion of people were involved in manufacturing, which was carried out in homes or small rural shops. Manufacturing was mostly done using hand tools and basic machines. A blacksmith created objects from iron by forging the metal. Furnaces were used to heat and soften the iron, and tools like hammers, anvils, and chisels were used to shape and create objects. A carpenter worked with wood and made household furniture, homes, wagons, tools, and even wooden utensils. The tools used were often made by the local blacksmith. Threads were spun by hand at home or in homes of spinners and weavers and made into cloth, hence the term "cottage industry." There were many other craftsmen , like tailors, jewelers, etc., who worked with their hands and were skilled at their work.

The **First Industrial Revolution** occurred in Britain at the end of the eighteenth century (roughly 1760–1840), though the precise start and end is debated by historians. The First Industrial Revolution saw the transition from skilled artisans making goods by hand to using machines for production. The Industrial Revolution started in and was largely confined to Britain, which at that time had great deposits of coal and iron ore. It was the leading power with colonies all over the world. Those colonies were sources of raw materials and consumers of the manufactured goods.

In the early part of the Industrial Revolution period, wind and water were the main sources of power for industries—waterwheels, windmills, and horsepower. This was replaced by steam power, with the development of stationary steam engine. Machines, such as lathes, planers, and milling and shaping machines, powered by steam engines were developed. Coal replaced wood and other biofuels as chief source of fuel in the iron industry. The spinning jenny was invented during this time and led to great development in the Textile industry. Chemicals were produced for the first time in large scale. Cement was patented and produced during this time. Agriculture also underwent mechanization in terms of development of mechanized plough, seed drill, and threshing machine. The transportation industry saw tremendous growth. Transport by land in Britain was mainly through waterways and road, both of which went through considerable improvements. The waterways are still in use in the United Kingdom. The development of the steam locomotive during this period marked the beginning of the development of railways.

The **Second Industrial Revolution** is attributed to the period from the end of nineteenth century to the beginning of the twentieth century, roughly from 1870 to 1914. This period saw rapid industrialization in Britain. Industrialization spread to Europe, in countries like France, Germany, and Belgium, and to the United States. The early twentieth century saw the United States emerge as a leading industrial power. Whereas the First Industrial Revolution saw the advent of many new technologies, the second revolution brought about many improvements in existing technologies. Transportation and telecommunications—railways and the telegraph system—were made available for use by the general public.

The **Second Industrial Revolution saw** the beginning of mass production, as well as the expanded use of electricity, petroleum, and steel. During this period, the first power station was established in Great Britain to supply public electricity. Prior to that, candles and gas lamps were used to light homes and factories. Electricity replaced water and steam in factories, which led to increased production. Steel replaced iron in construction projects, industrial machines, and many other areas. Steel made it possible to build public rail transport at competitive costs. This led to a surge in transportation, which in turn facilitated industrialization. Advancements in manufacturing, enabled developments such as the telegraph. This period saw extensive improvements to the water supply as well as sewage systems.

In America, after the Civil War (1861–1865), when the American colonies gained independence from Britain, the economy saw tremendous growth. America had abundance of natural resources and immigrants from Europe and Africa provided the workforce. The government actively promoted industrial development and land was granted to railroad companies to develop the railways, which in turn promoted mining of natural resources. The Industrial Revolution saw a tremendous growth in transportation—road, steamboats, railways—which help to link the communities that were previously isolated. There was rapid growth in factory production. The first car was built by Henry Ford, and the Ford Motor company used the concept of assembly lines to manufacture cars at affordable prices. The United States emerged as a leading industrial nation, outpacing Britain, France, and Germany.

The **First World War** (1914–1918) saw the end of the Second Industrial Revolution. Industrial development and other innovations were, however, used for military advantage. Mass production of guns and artillery and tanks was one such area. There was also rapid development of military aircraft used for reconnaissance, dropping bombs, and troop transport. Development of warships and submarines received a tremendous boost. The famous U-boats were built by Germany for the war.

The **Third Industrial Revolution**, also known as the **Digital Revolution**, commenced during the middle of the twentieth century. Digital electronics and digital computers ushered in the information age or information era. Digital computing and communication technology saw tremendous growth in this period. The merger of computers and communication technologies enabled access to large amounts of information in a timely manner and at low cost. The internet, another pivotal technology of this era, has become the backbone of networks and connectivity. The World Wide Web was introduced in early 1990s and soon the internet became a part of business and everyday life. It has transformed the way that people and businesses **connect with each other and access information**.

Information processing, information storage, and digital communications are some of the key outcomes of this era. Improvements in digital technologies include the growth of desktop computers to laptops to tablets. Landline phones have been replaced by mobile phones that have become smartphones. Communication technology has moved from 2G to 3G to 4G; 5G is on its way and will be soon available commercially. Information and knowledge have become the key to success and progress. The workforce has seen a tremendous change with a shift from the blue-collar factory worker to the white-collar information technology (IT) worker.

This revolution has impacted almost all sectors: manufacturing, healthcare, consumer finance, media, and many more industries. Our day-to-day business activities are automated and monitored using multiple devices like laptops, smartphones, and smart devices. The process of manufacturing has been transformed—whether in the supply chain, factory processes, marketing, sales, or research and development. Processes have been streamlined, errors minimized, quality has improved, and overall productivity has been enhanced.

We are now in the midst of the **Fourth Industrial Revolution.** Industry 4.0 combines traditional manufacturing and industry practices with technology to provide increased automation. It can be said to be the digital transformation of manufacturing. Automation is being taken to a new level with the introduction of cyber-physical systems. CPS are the integration of physical processes with networking and computation. Sensors are used to provide data input, **while embedded computers control the physical processes**. Machines, therefore, become independent entities that collect data, analyze data, and act upon the results in real time. They can operate independently and work with humans to create a cooperative environment. Smart factories or "Factories of the Future" is the keystone of Industry 4.0.

The other core technologies that are bringing about this transformation include the IoT, AI, big data analytics, and cloud computing. IoT is the technology that enables the interconnection of all types of devices through the internet. A network of sensors, devices, and actuators can exchange data and be monitored in real time. AI aims to mimic human intelligence to enable machines to make intelligent decisions. Machine leaning is an application of AI that enables machines to learn and improve from empirical/past data without being explicitly programmed to do so.

Big data describes huge volumes of structured, semi-structured, and unstructured data that can be analyzed/mined to obtain valuable insights. Factory automation data can be combined with data from other business processes, such as sales, marketing, and finance. The data can be processed and mined to obtain trends and other information, to improve production, business processes, and business in general.

Cloud computing is the delivery of different services through the internet, such as storage, databases, networking, software, and servers for doing the computing. Cloud service providers offer these services on-demand to customers over the internet. In other words, customers can requisition these services as and when they want. Small businesses benefit because they do not have to invest in an extensive computer infrastructure but instead pay only for the services they use. Large business can choose to invest in infrastructure for certain in-house computing and use cloud services for other business requirements. For example, big data processing can typically be accomplished using cloud services. Cloud providers host their services on servers located in different remote locations, hence the name cloud or virtual space.

Factories will therefore become increasingly automated and self-monitoring. Machines will have the ability to analyze information generated by them and other machines. Machines will be digitally connected with one another and will communicate with each other as well as with their human coworkers. Factories will become more efficient, productive, and economical.

Table 7.1 shows the different stages of the Industrial Revolutions along with the key developments in manufacturing during each period. The first Industrial Revolution used water and steam to mechanize production, the

TABLE 7.1

Stages of Industrial Revolution

Revolution	Key Developments in Manufacturing
First Industrial Revolution (1760–1840)	Mechanization Wind/water then steam power Coal as fuel
Second Industrial Revolution (1870–1914)	Mass production Assembly line Steel replaced iron Use of electricity and petroleum
Third Industrial Revolution (Middle of the twentieth century)	Use of computers for information processing Automation
Fourth Industrial Revolution (Emerging)	Cyber-physical systems (with its component technologies) Automation and interconnection

second revolution used electric power for mass production, and the third revolution used digital technology for computing and automation. The fourth uses **CPS**; working with a combination of technologies, to build the smart factories of the future.

7.2 Machine Vision and Industry 4.0

Machine vision is already an established component of automation in industry. Machine vision is a proven and established technology that is being used for various industrial applications. Industry 4.0, also known as IIOT (industrial IoT), is focused on connectivity and total factory automation. Smart equipment integrates the physical processes in a factory with networking and computing. Information exchange and processing enables real-time control of the physical process.

Embedded (smart) vision systems, composed of camera and processing unit, can be used in practically every step of the production process from procurement of raw material to delivery of the finished product to the customer. Thanks to rapid strides in vision technologies, such equipment is small and lightweight. These technologies can be used in the machines to monitor the production process and have the added advantage of being capable of deployment in difficult-to-access locations as well as locations hazardous to humans.

While quality assurance and control are important aspects of manufacturing, industry is poised to move from defect detection to defect prevention. Smart machines will use expert systems to control the manufacturing variables to improve the operations in real time based on feedback obtained from physical processes. However, when vision systems are extensively deployed for total factory automation, they will be capable of capturing all aspects of data pertaining to production. This will include process data, defects detected, related corrective action, and the tracking and use of raw material, works in progress, and finished products.

Special vision (image) processing software is being used currently to recognize objects and analyze images. Object recognition algorithms use databases with images of known objects to match with the images being processed to identify the objects. AI methods like Convolutional Neural Networks (CNN) are being increasingly used to train machines to identify and understand images. While traditional methods use software to recognize individual objects, deep learning approaches use generalized learning and training samples to recognize different classes of objects. The key advantage here is the capability for self-learning. The training samples must be large to obtain best results. The algorithms analyze and evaluate large amounts of training data to recognize patterns and subsequently use them for identifying the object. This greatly improves the accuracy.

Industry 4.0 factories will therefore increase production and productivity and enable the quick growth of business. The huge volume of production data accessed through vision equipment will be used to analyze and automatically flag defective products, enabling fast and accurate intervention to correct for errors such as defective inputs and/or process errors.

7.3 Emerging Vision Trends in Manufacturing

Defect detection and quality assurance will continue to be important applications. Intelligent systems and robust software will facilitate more accurate defect detection. For example, vision solutions are used for barcode and data codes identification. The technology will improve to recognize even blurry, distorted, or hard-to-read letters and numbers.

Use of mobile phones and other smart devices is likely to increase. More and more of the processing will likely be done with these smart devices and there will be less dependence on desktop computer or personal computers. Smart devices are likely to be equipped with high-resolution image cameras, copious RAM, and high-performance processors.

Supply chain management will see revolutionary changes. The connected supply chain will be able to accommodate new information as and when presented. For example, if a shipment of raw material is delayed, then the connected system will proactively modify manufacturing schedules to take care of such delays.

Autonomous vehicles and equipment are likely to be used for movement of supplies, goods, and personnel in the factory. Packing and dispatch of products will become highly automated. Automated logistics systems could assign the packed goods to outbound trucks and other transport carriers, like ships and aircraft, that will deliver the products to retailers as well as end-customers. Changes in arrival or departure schedules will be automatically adjusted based on inputs provided.

Robots are presently used only by large enterprises for a select segment of jobs. Robots will become more affordable for deployment even in smaller industries. The use of robots in factories will see a manifold increase. It is almost certain that robots will be used to carry out the mundane and repetitive activities as also activities that are hazardous to humans. Robots will be prevalent and involved in many autonomous activities, such as gripping, handling, and precisely placing parts; putting together subassemblies, assemblies, and final products; painting; quality checking; packaging and shipping; as well as in a variety of other manufacturing-related tasks.

Machine vision can also help in factory maintenance. Defective components in the embedded machines can be identified and replaced autonomously or by humans where required. Machines can also be involved in preventive maintenance. For example, a camera could detect problems in a work area or in machines, and human personnel can make the necessary rectifications. Likewise, if a smart machine detects that it is heating, it can shut down itself to avoid further damage.

Industry 4.0 is still in its nascent stage but is fast evolving into a comprehensive usable set of technologies. Large, medium, and small businesses and industries need to make a start in adopting these technologies. The changes to manufacturing methods and the environment as a whole are going to impact all manufacturers and it is crucial that they understand and implement the new practices. Medium and scale industries can leverage Industry 4.0 to improve their success rate. Technology can provide solutions to lower production costs and increase efficiency of services to provide a competitive edge. Technology can be leveraged to provide seamless, round-the-clock operations that efficiently and effectively meet all the customer requirements.

7.4 3D Imaging

We know that digital images and their processing are an essential part of machine vision applications. We are already familiar with 3D technology. We are getting used to 3D TVs, 3D movies, 3D computer monitors, 3D video games, etc. We now have 3D machine vision solutions. Why is the industry witnessing a sudden spurt in the application of 3D imaging for machine vision applications? Technological innovation has made the once expensive technologies significantly cheaper. The availability of cheaper 3D imaging cameras and sensors has seen the growth of new applications that use this technology. Applications that use 2D processing can only work with two dimensions of x and y, while 3D images provide length, width, as well as depth information, that is, the three dimensions x, y, and z. Nowadays, 3D imaging gives vision systems a new perspective—being able to obtain a realistic or replicated view of object in three dimensions. Some of the most common applications for 3D imaging include dimensional measurement, internal part analysis, and testing for consistency.

Two or more 2D images can also be used to obtain the third-dimension information. The geometrical structure of a scene is obtained from two or more images of the scene taken from different views. There are many ways of doing this. In stereo vision, images from two cameras are processed to obtain 3D information. The two cameras are displaced horizontally and capture two different views of the same object. The difference in the two images is used to obtain a disparity map that gives depth information. **Time-of-Flight (TOF)** scanning measures the time taken for light from a laser to travel between the camera and the object. In that way, 3D information such as depth as well as length and width of the object can be obtained. In 3D triangulation, a point in 3D is obtained given its projection (position) in two or more images. Finally, 3D image reconstruction is the process of creating a mathematical representation of a 3D image.

We know that 3D images provide depth and rotational information about the x-, y-, and z-axes. This additional information (as compared to 2D images) can be used for measurement, inspection, and positioning in quality control applications. For example, information related to shape features, such as object flatness, surface angle, and volume, can be obtained from 3D images. When contrast information is low or missing, 3D information can be used to detect parts and defects that have a measurable height difference. Text as well as code can also be read when contrast is low. Surface inspection methods are likely to significantly improve. It would become possible to identify tiny scratches or hairline cracks that are not visible to the human eye. Even unevenness in the machining may be detected. In fact, 3D machine vision methods may even make it possible to detect defects below the surface of objects, such as dust or air bubbles.

Technological advances have given rise to several new applications that provide accurate 3D information of objects. Structured light imaging obtains the 3D topography of objects by projecting a known pattern on the object scene. The object distorts the pattern, which is viewed using 2D cameras to obtain the depth and surface information. Structured light 3D scanners are noncontact 3D scanners that can be used in a variety of applications, for example, to lift fingerprints from a variety of surfaces. Police use this technique to process crime scenes.

Already, 3D X-ray imaging is a very important technology in the medical field, revolutionizing mammography and other kinds of medical imaging. We are already familiar with the use of CT scans for body scanning and diagnosis. Presently, 2D X-ray images captured from different angles can be used to reconstruct the three dimensions of an object. Since it is X-ray, it can also be used to obtain information about the inside of an object and can be used for defect analysis. 3D X-ray techniques are being progressively used for quality control of electronic equipment that are becoming more and more miniaturized and highly complex. For example, covered parts in printed circuit boards can be examined using 3D X-ray imaging, without having to take the object apart.

Another notable area of 3D imaging is in 3D printing, which is also called additive manufacturing. "Additive" refers to the successive addition of thin layers to create an object. Already, 3D printing is constructing layered objects for creating complex shapes. It can be used to create many different 3D printed objects that were previously only fabricated through mass manufacturing methods.

Another important application of 3D imaging is in the area of robotics and automation. For example, robots are used for bin picking, that is, locating parts from a bin and taking them out one by one as required. Robots require the components to be located or be positioned only in fixed spots, either in the bins as per example or elsewhere in the manufacturing process. Moveable parts need to be clamped down in a certain way for robots to always pick them up in the same way. Three-dimensional vision technologies can be adopted to enable robots to determine location so parts can be picked up even with variations. 3D vision technology can enable robots to behave like humans in plotting paths to avoid obstacles and prevent collusion and accidents. We will see the emergence of collaborative robots who will work with each other as well their human counterparts. The ultimate goal would be to use 3D imaging to allow robots to "see" the world just like humans do.

7.5 Emerging Vision Trends in Non-Manufacturing Applications

Vision solutions are being increasingly used in the nonmanufacturing sector. While the scope is enormous, some of the application areas include autonomous and driverless cars, agriculture, intelligent traffic systems, and medical imaging and diagnosis. We use the term "vision" here to refer to both computer vision and machine vision. Improvements in technology and emerging applications have caused computer vision and machine vision to seamlessly integrate and be commonly referred to as computer vision. Many or most of the vision applications process image data to enable real-time decision making in some form or another.

The construction industry has already made a start on automation. Robots are being used to lay bricks and build walls. A robot can lay bricks faster than a human without getting tired. This provides the double benefit of increased production with reduction in cost of construction. Robots can help with the heavy lifting, thereby reducing the burden on humans. Vision can help robots to navigate and avoid collisions. Resource tracking is another important area where vision solutions can be used. Tracking can be done from within construction sites or from remote locations. Automated tracking solutions can help to reduce costs and improve efficiency. What may have been taken up earlier as case studies will soon become reality.

We saw earlier that vision systems can construct 3D models from 2D images. Design visualization using 3D models helps designers and architects to visualize their ideas and identify potential problems. It speeds up the design process and provides a real view of how a city, factory, building, or house will look when finished—and all this, without using any of the actual construction resources like bricks, cement, labor, etc. It is particularly useful when many teams need to work together in large projects, like high-rise buildings and smart cities. For example, the finance team is able to know the overall cost and its allocation among building, transportation, labor, etc. Changes in a building can be made available as timely updates to all teams.

Customers or clients can also benefit from such 3D visualization of construction projects. Unlike 2D blueprints or drawings, 3D views can provide different and complete perspective of a house/building. 3D models can be animated so that clients can "walk through" their house and interact with the designers and architecture to suggest additions and modifications as per the preferences. Realistic lighting and furniture can enhance the viewing experience and help clients to fully understand what their final house would look like. 3D modeling can help clients to visualize not only the house but also its surroundings. Vision solutions and AI are likely to impact all aspects of construction—design, planning, building, safety, as well as monitoring and maintenance.

Computer vision is extensively used for medical imaging. Analysis of image data from different sources like X-rays, ultrasound, and other partial or full body scans are being used to provide better diagnosis and treatment. The vast data is also being used in the prediction of diseases. Early detection of life-threatening diseases like cancer can make the difference between life and death. Vision systems can be trained to detect symptoms that indicate the onset of such conditions or diseases. Hence, computer vision means not only providing vision but also reasoning skills to help the medical professionals in diagnosis. AI is expected to play a significant part in automatic learning and self-training systems for diagnosis.

Applications are being developed for sharing medical images and reports among radiologists, doctors, and other medical professionals. This helps to increase collaboration between medical professionals. Another related technology that may emerge would be the interconnection of medical imaging devices, making it possible to share devices across hospitals, health-care clinics, and vendors to do remote maintenance as well as preventive maintenance.

Robots are slowly but surely being extensively used in the field of medicine. Robots were initially used only for simple surgeries with minimal invasion. Robotic surgeries were performed by medical professionals in operation theatres. Computer vision adds the additional dimension of "vision" to robots. Present-day technology is so advanced that remote surgery and microrobot surgeries are being performed. The scope for the future is endless and will see the integration of medical imaging with IoT, AI and other technologies to improve diagnosis, treatment, medication, follow up, disease detection and prevention as well for research and collaborative work.

Robotics, or RPA (robotic process automation), is being used in many other fields to automate processes, most often repetitive processes, to increase performance throughput and efficiency. Vision can help robots to perform many operations—locating, sorting, sharing, moving, packing—and importantly collaborate to automate complex tasks.

It is only a matter of time before we see self-driving cars on the road and other autonomous vehicles being deployed for various activities. Self-driving cars have been proven to be feasible. However, they currently function only in predefined or protected environments. For example, a self-driving car may be able to comfortably drive itself only within a certain experimental campus. We have also seen self-driven locomotives that operate in certain sections but are yet to be used everywhere. Autonomous vehicles can avoid accidents that happen due to human errors and help to save lives.

Autonomous vehicles use devices like cameras, radar, and lasers, to "see" the world around them. Captured images need to be analyzed in real time to detect objects, which must be further classified and recognized in the context of the driving environment to make intelligent decisions. For example, is it an object on the road, like a stationary car, that needs to be avoided? Is it a slow-moving vehicle that needs to be overtaken? Is it perhaps a traffic signal that needs to be observed and traffic rules followed? The action would vary depending on the object and its location. Today AI is being extensively used to improve the accuracy of image analysis. One example is the use of CNN to classify and recognize objects. The advantage is that CNN are themselves self-learning and their ability to classify objects improves over time and use.

Another area where vision is applied in is tracking applications. For example, video surveillance is used to detect and track people, vehicles, and moving objects. Image content varies from video frame to frame because objects are moving due to camera motion, changes in the background, illumination variation, etc. In many applications, analysis needs to happen in real time. For example, unauthorized personnel would need to be detected and evicted from offices, buildings, schools, and other protected environments on a real-time basis. Again, AI is slated to be used to improve the accuracy and efficiency of the entire process.

Another emerging application is 3D mapping technology, that projects 2D or 3D objects onto a display surface using spatial mapping. This technology is also known as projection mapping. Projection mapping converts any surface into a dynamic visual display. It can be used to bring objects to life and can create an immersive environment. For example, 3D mapping can be used to add depth and movement to static objects like buildings. It can be used by artists and advertisers to add a new dimension to their work. It can be used for live concerts, in theaters, for decoration, etc. It will likely be used in the future by autonomous robots to explore unchartered terrains and space and convert their information into detailed 3D maps.

Readers may be aware of emerging technologies like augmented reality (AR) and virtual reality (VR). In AR, digital elements are added to a live view. For example, digital elements like people and other objects can be added to a normal water fountain to create an enthralling experience. It is possible to have virtual waterfalls and other animations in shopping malls to entertain shoppers. VR is a total digital experience that transports the user to a different world. It is an artificial computer-generated simulation of a real or imaginary world. For example, VR can transport you to places you have not seen but want to visit, like a penguin colony in Antarctica or maybe a journey deep into the ocean. Then we have merged reality (MR) where we can interact with the virtual worlds. Computer vision can add reality to the viewing experience. Cameras can be used to track the user. For example, eye tracking can be used to position the user in the virtual world. Details can be even added or rendered at the location where the person is looking.

7.6 Conclusion

Machine vision and robotics are already part of today's manufacturing world. Industry 4.0 has accelerated the move toward the fully automated factory. Factories of the future are expected to operate autonomously and collaborate with humans. Industry 4.0 integrates many technologies to create a connected manufacturing ecosystem that will boost productivity, enhance flexibility, reduce operating costs, and provide many more advantages for factories and industries. Machine vision will continue to be an essential element in automation. With Industry 4.0, vision systems can be expected to capture information about all aspects of production and delivery. This information can be intelligently and strategically leveraged for quality control and other applications. Data analytics and AI are likely to play crucial roles in making this happen.

The implementation and expansion of Industry 4.0 has created both challenges and opportunities for machine vision suppliers. We are likely to see a greater demand for machine vision systems. Machine vision software will see unexpected growth and integration with AI to enable deep learning in robotics and other processes and technologies. Machine vision is perceived to be complex and the need for ease of use will drive further standardization in machine vision products.

The lines between computer vision and machine vision have been blurring over the years as seen by the emerging use of vision application in the non-manufacturing sector. We can expect to see many more applications in the future, the limiting factor being perhaps only our imagination.

Industrialization has fundamentally transformed our environment and global ecosystem. We are in the era of the Fourth Industrial Revolution. Industry 4.0 integrates many powerful and transformative technologies that are developing at a faster pace than ever before. Our goal should be to avoid the mistakes of the past and ensure that these technologies are used for sustainable development that preserve our biodiversity.

Exercises

1. What is SLAM? Differentiate between GPS and SLAM.
2. What is a drone? Give examples of commercial applications of drones.
3. Is decentralization one of the design principles of Industry 4.0? Explain.
4. What is mass production? In which phase of the Industrial Revolution was mass production introduced? Discuss.
5. What is a smart factory? Explain.
6. What would be role of **IoT** in Industry 4.0?
7. Give an example of a medical robot in a hospital? Explain its functionality. Explain the use of computer machine vision in medical robots.
8. Discuss any emerging vision application in the entertainment industry.
9. Discuss an example of an quality control application that uses 3D image processing.
10. Explain the uses of computer/machine vision in automated monitoring of plants and trees.
11. Explain the difference between augmented reality and virtual reality. What is merged reality? Explain with examples.
12. Discuss in detail the use of augmented reality and virtual reality for a sample application in the fashion industry.

13. Discuss how you can use computer/machine vision to find the ingredients in a packaged food product.

14. What is point cloud technology? How is computer vision used in this technology?

15. Discuss how computer vision can be used for automated detection of cars that cross a tollbooth. Discuss uses of the information collected.

16. Explain with an example the role of robots (with computer vision) in packaging in industry.

17. Discuss how color image processing can be used to identify impurities in raw materials such as cotton fiber used for making cloth.

18. What is stereo imaging? Explain how it can used to obtain 3D images of objects.

19. What is occlusion in computer vision? What are the different types of occlusion? How can 3D imaging help in the recognition of occluded objects?

20. Write a note on research relating to the praying mantis (insect) and computer vision.

Bibliography

"A practical guide to machine vision lighting," January 30, 2017, http://www.ni.com/innovations-library/white-papers/.

Abrash, M 1997, "Chapter 35: Bresenham is fast, and fast is good," in: *Graphics Programming Black Book, Special Edition*, Coriolis Group Books, Washington, DC.

Adini, Y, Moses, Y & Ullman, S 1997, "Face recognition: The problem of compensating for changes illumination direction," *IEEE Transactions on Pattern Analysis and Machine*, vol. 19, pp. 721–732.

Agarwal, S & Roth, D 2002, "Learning a sparse representation for object detection," in: *European Conference on Computer Vision*, Springer, Heidelberg, vol. 4, pp. 113–127.

Ahearn, G 2015, "Digital inspection: A camera perspective," *Quality Magazine*.

Albahadily, HK, Tsviatkou, VYU & Kanapelka, VK 2017, "Grayscale image compression using bit plane slicing and developed RLE algorithms," *International Journal of Advanced Research in Computer and Communication Engineering*, vol. 6, issue 2. doi:10.17148/IJARCCE.2017.6272.

Alhamzi, K et al 2014, "3D object recognition based on image features: A survey," *International Journal of Computer and Information Technology*, vol. 3, pp. 651–660.

Artusi, A, Banterle, F, Ozan Aydın, T, Panozzo, D & Sorkine-Hornung, O 2016, *Image Content Retargeting: Maintaining Color, Tone, and Spatial Consistency*, CRC Press, Boca Raton, FL.

Barrow, H & Tenenbaum, J 1978, "Recovering intrinsic scene characteristics from images," in: *Computer Vision Systems*, Hanson, AR & Riseman EM (eds.), Academic Press, New York.

Bates, D 2017, Lens and sensor compatibility, White Paper, May 2017.

Bayona, A, SanMiguel, JC & Martínez, JM 2010, "Stationary foreground detection using background subtraction and temporal difference in video surveillance Alexander Hornberg 2006," *Handbook of Machine and Computer Vision*, Wiley.

Belhumeur, P, Hespanha, JP & Kriegman, DJ 1997, "Eigenfaces vs. Fisherfaces: Recognition using class specific linear projection," *IEEE Transactions on Pattern Analysis and Machine Intelligence*, vol. 19, issue 7, pp. 711–720.

Bergin, T 2010, Advantage of LED lighting in vision inspection systems, Quad Inc. White Paper.

Bo, L 2011, "Object recognition with hierarchical kernel descriptors," in: *Computer Vision and Pattern Recognition (CVPR), 2011 IEEE Conference*.

Bronstein, AM, Bronstein, M & Kimmel, M 2005, "Three-dimensional face recognition," *International Journal of Computer Vision*, vol. 64, issue 1, pp. 5–30.

Brooks, RA 1981, "Symbolic reasoning among 3-D models and 2-D images," *Artificial Intelligence Journal*, vol. 17, pp. 285–348.

Brown, M & Lowe, DG 2005, "Unsupervised 3d object recognition and reconstruction in unordered datasets," in: *International Conference on 3-D Digital Imaging and Modeling*, IEEE Computer Society, Washington, DC, pp. 56–63.

Buniatyan, D 2017, *Deep Learning Improves Template Matching by Normalized Cross Correlation Paperid 1705.08593*, Cornell University Library, New York.

Canny, J 1986, "A computational approach to edge detection," *IEEE Transactions on Pattern Analysis and Machine Intelligence*, vol. 8, issue 6, pp. 255–274.

Chandel, H & Vatta S 2015, "Occlusion detection and handling: A review," *International Journal of Computer Applications*, vol. 120, issue 10, pp. 33–38.

Chang, P & Krumm, J 1999, "Object recognition with color co-occurrence histogram," in: *IEEE Conference on CVPR 1999*, Fort Collins, CO.

Cipolla, R, Hernandez, C, Vogiatzis, G & Stenger, B 2007, "Recent advances and new applications," *Computer Vision, Special Reports*, Computer vision group, vol. 62, issue 12, pp. 62–68.

Clowes, MB 1971, "On seeing things," *Artificial Intelligence*, vol. 2, pp. 79–116.

Collet Romea, A, Berenson, D, Srinivasa, S & Ferguson, D 2009, "Object recognition and full pose registration from a single image for robotic manipulation," in: *Proceedings of the ICRA*, Kobe, Japan.

Comaniciu, D & Meer, P 2002, "Mean shift: A robust approach toward feature space analysis," *IEEE Transactions on Pattern Analysis and Machine Intelligence*, vol. 24, pp. 603–619.

Costa, M 2000, "3D object recognition and pose with relational indexing," *Computer Vision and Image Understanding*, vol. 79, pp. 364–407.

Crimson, W 1980, "A computer implementation of a theory of human stereo vision," AI Memo 565, MIT.

Cubero, S, Aleixos, N, Moltó, E, Gómez-Sanchis, J & Blasco, J 2011, "Advances in machine vision applications for automatic inspection and quality evaluation of fruits and vegetables," *Food Bioprocess Technology*, vol. 4, pp. 487–504. doi:10.1007/s11947-010-0411-8.

Dai, Q & Hoiem, D 2012, "Learning to localize detected objects," in: *IEEE Conferences on Computer Vision and Pattern Recognition*.

Das S, Saikia, J, Das, S & Goni, N 2015, "A comparative study of different noise filtering Techniques in digital images," *International Journal of Engineering Research and General Science*, vol 3, issue 5, pp. 180–190.

Davies, ER 1998, "Automated visual inspection," in: *Machine Vision*, Chapter 19, 2nd edn, Academic Press, pp. 471–502.

Dhawanm, AP 2003, *Medical Image Analysis*, IEEE Press-Wiley, Hoboken, NJ.

Duda, R & Hart, P 1972, "Use of the Hough transformation to detect lines and curves in pictures," *Communications of the ACM*, vol. 15, issue 1, pp. 11–15.

Fadnavis, S 2014, "Image interpolation techniques in digital image processing. An Overview," *International Journal of Engineering Research and Applications*, vol. 4, issue 10, pp. 70–73.

Fairchild, MD 2013, *Color Appearance Models*, 3rd edn, Wiley Publishers, Chichester, UK.

Felzenszwalb, P, McAllester, D & Ramanan, D 2008, "A discriminatively trained, multiscale, deformable part model," in: *IEEE Conference on Computer Vision and Pattern Recognition*, University of Chicago, Chicago, IL.

Fergus, R, Perona, P & Zisserman, A 2003, "Object class recognition by unsupervised scale-invariant learning," in: *IEEE Computer Society Conference on Computer Vision and Pattern Recognition*, IEEE.

Finlayson, G, Hordley, S & Morovic, P 2005, "Colour constancy using the chromagenic constraint," in: *IEEE Conference on Computer Vision and Pattern Recognition*, pp. 1079–1086.

Fleming, RW 2014, "Visual perception of materials and their properties," *Vision Research*, vol. 94, pp. 62–75.

Freeman, WT 2011, "Where computer vision needs help from computer science," *ACM-SIAM Symposium on Discrete Algorithms (SODA)*, Computer Science and Artificial Intelligence, MIT, Cambridge, MA.

Gavrila, DM 1999, "The visual analysis of human movement, a survey," *Computer Vision and Image Understanding*, vol. 73, issue 1, pp. 82–98.

Gavrilova, ML & Monwa, M 2013, *Multimodal Biometrics and Intelligent Image Processing for Security Systems*, Information Science Reference, Hershey, PA.

Gernsheim, H 1977, "The 150th anniversary of photography," *History of Photography*, vol. I, issue 1, pp. 3–8.

Goad, C 1986, "Special purpose automatic programming for 3D model-based vision," in: *From Pixels to Predicates*, Ablex Publishing, Norwood, NJ, pp. 371–391.

Gokmen, O, Acar, C, Arribas-Lorenzo, G & Morales, FJ 2008, "Investigating the correlation between acrylamide content and browning ratio of model cookies," *Journal of Food Engineering*, vol. 87, issue 3, pp. 380–385.

Gonzalez, RC & Woods, RE 2002, *Digital Image Processing*, 2nd edn, Addison-Wesley, Reading, MA.

Goshtasby, AA 2012, *Image Registration Advances in Computer Vision and Pattern Recognition*, Springer-Verlag, London, UK.

Grauman, K & Leibe, B 2011, *Visual Object Recognition*, Morgan & Claypool Publishers, San Rafael, CA.

Gray, A 1997, *Euclidean Spaces in Modern Differential Geometry of Curves and Surfaces with Mathematica*, 2nd edn, CRC Press, Boca Raton, FL, pp. 2–5.

Grimson, WEL 1988, "The combinatorics of object recognition in cluttered environments using constrained search," in: *Second International Conference on Computer Vision, IEEE*, pp. 218–227.

Guzman, A 1968, "Computer recognition of three-dimensional objects in a visual scene," Technical Report MAC-TR-59 MIT, Cambridge, UK.

Hager, GD 2003, "Problems of computer vision: Recognition," in: *IEEE International Conference on Robotics and Automation*.

Haralick, R & Shapiro, L 1992, *Computer and Robot Vision*, Addison-Wesley Longman Publishing Co., Boston, MA.

Haralick, RM 1984, "Digital step edges from zero crossing of second directional derivatives," *IEEE Transactions on Pattern Analysis and Machine Intelligence*, vol. PAMI-6, issue 1, pp. 58–68.

Hartley, RI & Zisserman, A 2000, *Multiple View Geometry in Computer Vision*, Cambridge University Press, Cambridge, UK.

Hassaballah, M, Abdelmgeid, AA & Alshazly, HA 2016, "Image features detection, description and matching," *Studies in Computational Intelligence*, vol. 630, pp. 11–45. doi:10.1007/978-3-319-28854-3_2.

Hissmann, O 2004, "Film inspection—A question of location," *Kunststoffe Plast Europe*, vol. 6, pp. 76–78.

Horn, BKP 1975, "Obtaining shape from shading information," in: *The Psychology of Computer Vision*, Winston, PH (ed.), McGraw-Hill, New York, p. 115.

Horn, BKP 2000, "Tsai's camera calibration method revisited." http://mit.sustech.edu/.

Horprasert, T, Harwood, D & Davis, LS 1999, "A statistical approach for realtime robust background subtraction and shadow detection," in: *Proceedings of the IEEE Frame Rate Workshop*, Kerkyra, Greece, pp. 1–19.

Hsiao, E, Collet, D & Hebert, M 2010, "Making specific features less discriminative to improve point-based 3d object recognition," in: *Proceedings of the IEEE Computer Society Conference on Computer Vision and Pattern Recognition*, San Francisco, CA.

http://multiplex.com

https://www.edmundoptics.com/resources/application-notes/imaging/

Huang, TS 1996, "Computer vision: Evolution and promise," in: *International Conference: 5th, High Technology: Imaging Science and Technology*, Chiba, Japan.

Huttenlocher, D & Ullman, S 1990, "Recognizing solid objects by alignment with an image," *International Journal of Computer Vision*, vol. 5, issue 2, pp. 195–212.

"Imaging electronics 101: Understanding camera sensors for machine vision applications," https://www.edmundoptics.com/resources/application-notes/imaging/understanding-camera-sensors-for-machine-vision-applications/.

Jain, T & Meenu 2013, "Automation and integration of industries through computer vision systems," *International Journal of Information and Computation Technology*, vol. 3, issue 9, pp. 963–970.

Jean, P 2014, *Computer Vision: A Modern Approach*, 2nd edn, Prentice-Hall, Upper Saddle River, NJ.

Kalirajan, K & Sudha, M 2015, "Moving object detection for video surveillance," *The Scientific World Journal*, vol. 20, Article ID 907469.

Kaur, D et al 2014, "Various image segmentation techniques: A review," *International Journal of Computer Science and Mobile Computing*, vol. 3, issue 5, pp. 809–814.

Kepf, P 2016, "Seven important factors for selecting machine vision lens," computar, computar.com.

Kerlin, F 1988, "Meaningfulness and the Erlanger program of Felix Klein," *Mathématiques, Informatique et Sciences Humaines*, vol. 26, issue 101, pp. 61–71.

Khurana, K et al 2013, "Techniques for object recognition in images and multi-object detection," *International Journal of Advanced Research in Computer Engineering & Technology (IJARCET)*, vol. 2, issue 4, pp. 1383–1388.

Kushal, A, Schmid, C & Ponce, J 2007, "Flexible object models for category-level 3D object recognition," in: *Proceedings of IEEE International Conference on Computer Vision and Pattern Recognition*. doi:10.1109/CVPR.2007.383149.

Lampert, C, Blaschko, M & Hofmann, T 2008, "Beyond sliding windows: Object localization by efficient subwindow search," in: *IEEE Conference on Computer Vision and Pattern Recognition*, Anchorage, AK.

Latharani, TR et al 2011, "Various object recognition techniques for computer vision," *Journal of Analysis and Computation*, vol. 7, issue 1, pp. 39–47.

Lazebnik, S, Schmid, C & Ponce, J 2004, "Semi-local affine parts for object recognition," in: *British Machine Vision Conference (BMVC '04)*, September, Kingston, UK, pp. 779–788.

Leibe, B, Leonardis, A & Schiele, B 2008, "Robust object detection with interleaved categorization and segmentation," *IJCV*, vol. 77, pp. 259–289.

Levine, MD & Bhattacharyya, J 2005, "Removing shadows," *Pattern Recognition Letters*, vol. 26, pp. 251–265.

Liao, H, Edwards, P, Pan, X, Fan, Y & Yang, GZ 2010, *Medical Imaging and Augmented Reality*, vol. 6326 of Lecture Notes in Computer Science, Springer, Berlin, Germany, pp. 372–383.

Lowe, DG 1987, "Three-dimensional object recognition from single two-dimensional images," *Artificial Intelligence*, vol. 31, issue 3, pp. 355–395.

Lowe, DG 1999, "Object recognition from local scale-invariant features," in: *ICCV.*

Lowe, DG 2001, "Local feature view clustering for 3D object recognition," in: *IEEE Computer Society Conference on Computer Vision and Pattern Recognition CVPR,* Madison, WI.

Lowe DGb & Scott 2009, "The Tech Behind REGION," *Proceedings of the IEEE Conference on Computer Vision and Pattern Recognition,* December 2001, Kauai, HI, pp. 1–7.

Malcolm, D 2004, "William Henry Fox Talbot (1800–1877) and the Invention of Photography," in: *Heilbrunn Timeline of Art History,* The Metropolitan Museum of Art, New York.

Marr, D 1982, *VISION a Computational Investigation into the Human Representation and Processing of Visual Information,* W. H. Freeman and Company, San Francisco, CA.

Martin, D 2007, Practical guide to machine vision lighting, Rauscher, pp. 1721–1728, http://ieeexplore.ieee.org/stamp/stamp.jsp?tp=&arnumber=7080207&isnumber=7079654.

Matas, J 2004, "Object recognition methods based on transformation covariant features," in: *Proceedings of European Signal Processing Conference on EUSIPCO,* Vienna, Austria

Mikolajczyk, K & Schmid, C 2004, "Scale and Affine invariant interest point detectors," *International Journal of Computer Vision,* vol. 60, issue 1, pp. 63–86.

Mohammed Yasin, AS, Haque M, Binte Anwar, S, & Ahamed Shohag, S 2013, "Computer vision techniques for supporting blind or vision impaired people: An overview," *International Journal of Scientific Research Engineering & Technology (IJSRET),* vol. 2, issue 8, pp. 498–503.

Mundy, JL 2006, "Object recognition in the geometric era: A retrospective," in: *Toward Category-Level Object Recognition,* vol. 4170 of Lecture Notes in Computer Science, Springer, pp. 3–28.

Naik, S & Patel, B 2017, "Machine vision based fruit classification and grading—A review," *International Journal of Computer Applications (0975 – 8887),* vol. 170, issue 9, pp. 22–32.

Nalwa, VS & Binford, TO 1986, "On detecting edges," *IEEE Transactions on Pattern Analysis and Machine Intelligence,* vol. 8, pp. 699–714.

Neugebauer, PJ 1997, "Reconstruction of real world objects via simultaneous registration and robust combination of multiple range images," *International Journal of Shape Modeling,* vol. 3, issue 1–2, pp. 71–90.

Nilsson, NJ 2009, *The Quest for Artificial Intelligence,* Cambridge University Press, Cambridge, UK.

Nister, D & Stewenius, H 2006, "Scalable recognition with a vocabulary tree center for visualization and virtual environments," Department of Computer Science, University of Kentucky. doi:10.1.1.61.9520.

Novak, A 2008, Quillan leaf controversy, "Does Sotheby's go overboard with its auction promotion? Is photo history the victim?" issue 148. http://www. iphoto-central.com/news/article_view.php/157/148/877.

Okabe, T, Kondo, Y, Kitani, KM & Sato, Y 2010, "Recognizing multiple objects based on co-occurrence of categories," Special issue: 3D image and video technology, *Progress in Informatics,* issue 7, pp. 43–52.

Parraga-Alava, JA 2015, "Computer vision and medical image processing: A brief survey of application areas," *ASAI,* 16° 44 JAIIO, pp. 152–159.

Pavan Kumar, G & Bhanu Prasad, P, "Machine vision based quality control: Importance in pharmaceutical industry," *International Journal of Computer Applications (0975 – 8887), International Conference on Information and Communication Technologies (ICICT-2014)*, vol. 0975–8887, pp. 30–36.

Pearson, TC, Moore, D & Pearson, JE 2011, "A machine vision system for high speed sorting of small spots on grains," *Sensors and Instrumentation for Food Quality and Safety*, vol. 3, pp. 5–8.

Pesco, DU 2001, Matrices and digital images. Institute of Mathematics and Statistics Fluminense Federal University.

Pinto, N, Cox, D & DiCarlo, J 2008, "Why is real-world visual object recognition hard?" *PLOS Computational Biology*, vol. 4, issue 1, p. E27.

Remondino, F & Fraser, C 2006, "Digital camera calibration methods: Considerations and comparisons," in *ISPRS Commission V Symposium Image Engineering and Vision Metrology*, pp. 266–272.

Riesenhuber, M & Poggio, T 1999, in: *Biologically Motivated Computer Vision*, Lee, S-W, Bulthoff, HH & Poggio, T (eds.), Nature Neuroscience, Washington, DC, pp. 1–9.

Robb, RA 2000, *Biomedical Imaging, Visualization and Analysis*, Wiley Publications, New York.

Roberts, L 1965, "Machine perception of 3-D solids," *OEOIP,* , vol. 91, issue 3, pp. 159–197.

Roth, M, Tanaka, K, Weissman, C & Yerazunis, W 1999, *Computer Vision for Interactive Computer Graphics*, Mitsubishi Electric Research Laboratories, Cambridge, MA.

Rothganger, F, Lazebnik, S, Schmid, C & Ponce, J 2006, "3D object modeling and recognition using local affine-invariant image descriptors and multi-view spatial constraints," *International Journal of Computer Vision*, vol. 66, issue 3, pp. 231–259.

Ruikar, SD 2011, "Wavelet based image denoising technique," *International Journal of Advanced Computer Science and Applications*, vol. 2, issue 3, pp. 49–53.

Sachdeva, M, Dharni, N & Mehta, K 2014, "Computer vision," *International Journal for Research in Applied Science and Engineering Technology*, vol. 2, issue IX, pp. 1–4.

Safinaz, S 2014, "An efficient algorithm for image scaling with high boost filtering," *International Journal of Scientific and Research Publications*, vol. 4, issue 5, pp. 1–9.

Scharstein, D & Szeliski, R 1998, "Stereo matching with nonlinear diffusion," *International Journal of Computer Vision*, vol. 28, issue 2, pp. 155–174.

Sebe, N & Lew, MS 2013, *Robust Computer Vision: Theory and Applications*, Springer Publications, Basel, Switzerland.

Selinger, A & Nelson, RC 1999, "A perceptual grouping hierarchy for appearance-based 3D object recognition," *Computer Vision and Image Understanding*, vol. 76, issue 1, pp. 83–92.

Shapiro, L & Stockman, G 2001, *Computer Vision*, Prentice-Hall, Upper Saddle River, NJ.

Shashidhar, V, Aruna Kumara, B, Neelu, L & Bharath, J 2015, "Digital image classification and clustering," *International Journal of Innovative Research in Computer and Communication Engineering*, vol. 3, issue 5, pp. 4187–4197.

Siswantoro, J, Prabuwono, AS, Abdullah, A & Bahari, I 2014, "Monte Carlo method with heuristic adjustment for irregularly shaped food product volume measurement. Research Article," *The Scientific World Journal*. doi:10.1155/2014/683048.

Sklansky, J 1978, "On the Hough technique for curve detection," *IEEE Transactions on Computers*, vol. C-27, issue 10, pp. 923–926.

Struwe, M 2016, *Appearance-based Object Recognition Models*, Faculty of Technology, Bielefeld University, Bielefeld, Germany.

Sturm, P, Ramalingam, S, Tardif, J-P, Gasparini, S & Barreto, J 2011, "Camera models and fundamental concepts used in geometric computer vision," *Foundations and Trends in Computer Graphics and Vision*, vol. 6, issue 1–2, pp. 1–183.

Suzuki, MT, Yaginuma, Y & Shimizu, Y 2005, "A partial shape matching technique for 3D model retrieval systems," *Proceeding SIGGRAPH '05 ACM SIGGRAPH*, Los Angeles, CA.

Swinton, AAC 2017, *The Elementary Principles of Electric Lighting*, Reprint by ReInk Books.

Szeliski, R 2010, *Computer Vision: Algorithms and Applications*, Springer Science & Business Media, Berlin, Germany, pp. 10–16.

Tiwar, US 2009, *Proceedings of the First International Conference on Intelligent Human Computer Interaction*, Springer Publications, Orlando, FL.

Toriwaki, J & Yoshida, H 2009, *Fundamentals of Three Dimensional Image Processing*, Springer Science & Business Media, New York.

Ulrich, M & Steger, C 2001, "Empirical performance evaluation of object recognition methods," in: Christensen, HI & Phillips, PJ (eds.), *Empirical Evaluation Methods in Computer Vision*, IEEE Computer Society Press, Los Alamitos, CA, pp. 62–76.

Vrigkas, M, Nikou, C & Kakadiaris, IA 2015, "A review of human activity recognition methods," *Frontiers in Robotics and AI*. doi:10.3389/FROBT.2015.00028.

Waltz, DL 1972, "Generating semantic descriptions from drawings of scenes with shadows," MIT internal report, reprinted as "Understanding Line Drawings of Scenes with Shadows," in: *The Psychology of Computer Vision*, vol. 19, issue 92. Winston, PH (ed.), McGraw-Hill, New York.

Weber, M, Welling, M & Perona, P 2000, "Towards automatic discovery of object categories," in: *IEEE Conference on Computer Vision and Pattern Recognition*, Hilton Head Island, SC.

Wu, C, Clipp, B, Li, X, Frahm, J-M & Pollefeys, M 2011, "3D model matching with Viewpoint-Invariant Patches (VIP)," Department of Computer Science, The University of North Carolina, Chapel Hill, NC.

Wu, Y 2001, "Vision and learning for intelligent human-computer interaction," PhD Thesis, University of Illinois at Urbana-Champaign, Champaign, IL.

www.agrilabour.com

www.brainpickings.org

www.visiononline.org/

Zerroug, M & Nevatia, R 1996, "Three-dimensional descriptions based on the analysis of the invariant and quasi-invariant properties of some curved-axis generalized cylinders," *IEEE Transactions on Pattern Analysis and Machine Intelligence*, vol. 18, issue 3, pp. 237–253.

Zhang, S, Qu, X, Ma, S, Yang, Z & Kong, L 2012, "A dense stereo matching algorithm based on triangulation," *Journal of Computational Information Systems*, vol. 3, pp. 283–292.

Zhao, WH & Chellappa, R 2000, "Illumination insensitive face recognition using symmetric shape-from shading," *IEEE Conference on Computer Vision and Pattern Recognition*, vol. 1, pp. 286–293.

Zheng, L, Li, G & Sha, J 2007, "The survey of medical image 3D reconstruction," in: *Proceedings of the SPIE 6534, Fifth International Conference on Photonics and Imaging in Biology and Medicine*. doi:10.1117/12.741321.

Zimmerman, G, Legge, G & Cavanagh, P 1995, "Pictorial depth cues: a new slant," *Journal of the Optical Society of America A*, vol. 12, issue 1, pp. 17–26.

Index